574.88
F88m

114083

DATE DUE			
Dec3 30			

A SERIES OF BOOKS IN BIOLOGY
Cedric I. Davern, EDITOR

MOLECULAR BIOLOGY AND BIOCHEMISTRY

Problems and Applications

David Freifelder

BRANDEIS UNIVERSITY

W. H. FREEMAN AND COMPANY

San Francisco

Library of Congress Cataloging in Publication Data

Freifelder, David Michael, 1935–
　　Molecular biology and biochemistry: Problems and applications.

　(A Series of books in biology)
　Includes bibliographies and indexes.
　　1.　Molecular biology.　2.　Molecular biology—
Problems, exercises, etc.　I.　Title.
QH506.F73　　　　　574.8'8　　　　　78-18712
ISBN 0-7167-0068-9 pbk.

Contents

Preface

In any experimental science it is important that students have practice in dealing with experimental observations. Carefully selected problems and questions can give students such practice; furthermore, these give a student an opportunity to test his or her understanding of textbook and lecture material and to put together various facts in order to draw conclusions not explicitly stated elsewhere. In the past, in my own course in molecular biology, I provided homework problems which I myself wrote, since the existing textbooks contained few problems. Writing problems for homework is a time-consuming operation compounded by the worry that a problem might be thought ambiguous or be misconstrued by the student who perceives details more sharply than concepts or, even worse, might not be answerable. All teachers face this problem but, after several years, usually accumulate a set of unambiguous, solvable problems.

In an effort to provide students and teachers of molecular biology, genetics, and biochemistry with useful problems and questions, I have collected homework sets and examinations compiled by several well-known molecular biologists who have been teaching—and assigning problems—for many years. These sets of problems form the core of this book. They range in difficulty from medium to challenging. As I arranged the material for this book, I felt it best to add about 300 relatively simple problems and questions for the beginning student.

It is intended that this book be used with the most widely used texts in molecular biology, genetics, and biochemistry. Since the chapter headings I have used in this book do not match those of every text, I have included lists at the front of the book of the problems appropriate for chapters in the major texts. Furthermore, an index is provided in which the problems are sorted by topic.

How to Use This Book

Molecular Biology and Biochemistry is meant to be a study guide, to teach as well as to test one's understanding. It is assumed that students using this book are taking a course in molecular biology or genetics and are reading an adequate textbook.

Each chapter deals with a separate subject and begins with several pages of explanatory material. These introductions are not meant to replace a textbook, and some cover only some of the topics within the chapters. Their purpose is to summarize basic material, define terms that must be known to solve the problems, provide some material that frequently is not contained in the usual textbooks, and to restate those points experience has shown students in molecular biology courses often miss or forget.

Students who supplement the introductions with the literature referenced at the end of each introduction should be able to solve all the problems.

The problems marked with an open circle (o) are elementary (though not necessarily simple) and are included so that students can be sure they understand various terms and know basic facts and concepts. The elementary problems should be answered first. If unable to answer any of these, a student should not proceed further until having done some review.

The problems marked with a solid circle (●) are difficult and often are beyond the range of a beginning course.

Answers to selected problems—those designated with boldface numbers—are given at the end of the book. Inasmuch as the problems have been grouped by subject, when several problems deal with the same subject, generally only one is answered. Similarly, if a problem has several parts which differ only slightly, only a few parts are answered. Answers are usually not as brief as a single word, except for very simple problems; in general, the answer is in the form of an explanation, which should itself be a teaching guide.

The section "Problems Useful for Particular Textbooks" at the beginning of the book and the "Index of Problems" at the end are intended to help students help themselves to a great extent: the former offers a convenient way to locate problems pertinent for use with various texts, and the latter is organized by concept so that (1) if a student wishes to test his or her ability in a given subject, appropriate problems can be located, and (2) if a difficult problem is encountered and one thinks it deals with a particular topic, a student can look up that topic to confirm that it is indeed applicable to the problem being solved.

I would like to express thanks to Bruce Alberts, Rich Calendar, Hatch Echols, Julie Marmur, Frank Stahl, and Bob Warner for supplying problems, to many of my students for reading the problems and reporting difficulties they experienced while solving them, to Ric Davern, who suggested that I gather the best problems of experienced molecular biologists, to Phil Hanawalt, who suggested that a problems book would be more useful if it were also an informative study guide, to Jay Magno, for the art work, and to Mildred Kravitz and Barbara Nagy, who typed the many drafts of the manuscript.

Waltham, Massachusetts   *David Freifelder*
July 1978

Problems Useful for
Particular Textbooks

In the following lists, beneath each textbook citation, "chapter" refers to chapters within that textbook and "problems" refer to problems within this book.

Davis, B. D., R. Dulbecco, H. N. Eisen, H. S. Ginsberg, and W. B. Wood. 1973. *Microbiology*. New York. Harper & Row.

Chapter	Problems
9	2-2 to 2-6; 17-1 to 17-24, 17-26, 17-35, 17-40, 17-42
10	4-1 to 4-12, 4-18 to 4-61; 5-1 to 5-33, 5-35, 5-39 to 5-47; 7-1
11	1-24 to 1-27; 5-34; 7-1 to 7-13, 7-16, 7-18, 7-20 to 7-34; 8-1 to 8-15, 8-51 to 8-54
12	7-13, 7-23; 9-1 to 9-10; 10-1 to 10-25; 11-1 to 11-17
13	8-16 to 8-36, 8-39 to 8-55; 12-1 to 12-17
44	14-27; 17-18 to 17-28
45	15-1 to 15-44, 15-79, 15-80; 16-3, 16-31 to 16-33, 16-44 to 16-50
46	15-45 to 15-72, 15-76, 15-77, 15-81 to 15-85; 16-4 to 16-6, 16-30, 16-36 to 16-38, 16-41
47	1-1 to 1-5, 1-17

Kornberg, A. 1974. *DNA Synthesis*. San Francisco. W. H. Freeman and Company.

Chapter	Problems
1	4-1 to 4-72
4	5-1, 5-2, 5-17, 5-24; 6-1, 6-2; 7-1, 7-3, 7-5, 7-23
5	5-1, 5-2, 5-4, 5-17
6	6-13; 16-15
7	5-1 to 5-47
8	5-3; 6-3; 15-27, 15-28, 15-29
9	6-5, 6-16; 13-1 to 13-13; 16-44, 16-45, 16-47, 16-48, 16-50
10	8-1 to 8-56
11	6-15

Lehninger, A. L. 1974. *Biochemistry*. New York. Worth Publishers, Inc.

Chapter	Problems*
5	3-1 to 3-10
6	3-11 to 3-24
9	12-1 to 12-17
31	2-1 to 2-16; 4-1 to 4-73; 7-1 to 7-33
32	5-1 to 5-47; 6-1 to 6-17; 13-1 to 13-13
33	8-1 to 8-16; 10-1 to 10-25
34	9-1 to 9-21; 11-1 to 11-17
35	8-17 to 8-56

* Many of these problems are at a more advanced level of molecular biology than the sections in Lehninger. They are included because many could be worked if supplementary information were given by the instructor.

Lewin, B. 1974. *Gene Expression, Volume I*. New York. Wiley-Interscience.

Chapter	Problems
1	7-6, 7-8, 7-10, 7-18, 7-28; 9-1 to 9-19; 11-1 to 11-17
2	10-1 to 10-25
3	7-7, 7-17; 8-38; 10-1 to 10-25
5	9-1 to 9-19; 10-1 to 10-25; 11-1 to 11-17; 15-35, 15-37, 15-41 to 15-44, 15-48
6	8-1 to 8-56; 15-46
7	8-19, 8-20, 8-23, 8-26, 8-28 to 8-35, 8-38 to 8-40
8	8-37; 15-45 to 15-68, 15-70; 16-8, 16-9
10	5-1 to 5-47; 6-1 to 6-17; 7-1, 7-4, 7-23
11	13-1 to 13-12
12	16-1 to 16-50
13	5-19, 5-37 to 5-40

Stryer, L. 1975. *Biochemistry*. San Francisco. W. H. Freeman and Company.

Chapter	Problems
2	3-1 to 3-24
6	12-1 to 12-17
23	2-1 to 2-6; 4-1 to 4-72; 5-1 to 5-47; 6-4
24	8-1 to 8-10, 8-13, 8-14, 8-52 to 8-56
25	7-1 to 7-32; 9-1 to 9-20; 10-1 to 10-25
26	8-15; 10-1 to 10-25; 11-1 to 11-17; 15-20
27	8-11, 8-16 to 8-51
29	6-5, 6-13; 15-1, 15-27 to 15-31, 15-43, 15-45, 15-46, 15-55; 17-1 to 17-28

Stent, G., and R. Calendar. 1978. *Molecular Genetics* Second Edition, San Francisco. W. H. Freeman and Company.

Chapter	Problems
1	1-17 to 1-27
2	1-6 to 1-9
3	1-10 to 1-16
4	3-1 to 3-24; 12-1 to 12-17
5	1-17 to 1-21; 7-19; 8-5
6	5-34; 7-1 to 7-13, 7-18, 7-20 to 7-34; 11-1 to 11-17
7	2-2 to 2-4; 5-35
8	4-1 to 4-61; 5-1 to 5-47; 6-1 to 6-17; 7-1, 7-13; 9-1 to 9-20
9	16-1, 16-2, 16-6 to 16-15, 16-24, 16-35, 16-42
10	15-1 to 15-97; 16-5, 16-25, 16-27, 16-31, 16-34, 16-36 to 16-38
11	1-24 to 1-27; 16-26 to 16-30, 16-44 to 16-50
12	5-34; 7-1 to 7-13, 7-18, 7-20 to 7-34; 9-1 to 9-20
13	16-4, 16-5, 16-20, 16-40
14	7-16; 8-1 to 8-15, 8-51 to 8-55
15	10-1 to 10-25
16	7-13; 9-1 to 9-20
17	None
18	5-1 to 5-47
19	13-1 to 13-13
20	5-37, 5-38; 8-16 to 8-36, 8-39 to 8-53; 15-46 to 15-51
21	None

Suzuki, D. T., and A. J. F. Griffiths. 1976. *An Introduction to Genetic Analysis*. San Francisco. W. H. Freeman and Company.

Chapter	Problems
1	1-17 to 1-22
4	1-23 to 1-27
6	16-1 to 16-9
8	2-2 to 2-4; 16-1, 16-2
10	2-5, 2-6; 4-1 to 4-4, 4-6 to 4-9, 4-17, 4-19 to 4-21, 4-34; 5-5 to 5-8, 5-31, 5-32, 5-37, 5-38, 5-42; 11-1 to 11-17
11	9-1 to 9-20; 10-1 to 10-25
12	7-1 to 7-33
13	8-17 to 8-23, 8-27 to 8-31, 8-33 to 8-35.

Watson, J. D. 1976. *Molecular Biology of the Gene*. Third Edition. Menlo Park, Calif., W. A. Benjamin.

Chapter	Problems
1	1-1 to 1-6; 17-2 to 17-7
2	1-10 to 1-16; 2-2 to 2-6; 4-3
3	None
4	3-1 to 3-24; 4-6 to 4-11, 4-18 to 4-20, 4-23 to 4-30
5	None
6	3-1 to 3-34
7	1-24 to 1-27; 15-1 to 15-11; 16-1 to 16-27, 16-31, 16-34, 16-35, 16-42
8	1-17 to 1-27; 7-19; 8-5
9	4-1 to 4-61; 5-1 to 5-33, 5-35, 5-39 to 5-47; 7-1
10	5-34; 7-1 to 7-13, 7-18, 7-20 to 7-34
11	7-16; 8-1 to 8-15, 8-51 to 8-54
12	10-1 to 10-25
13	7-13, 7-23; 9-1 to 9-20; 11-1 to 11-17
14	8-16 to 8-36, 8-39 to 8-55
15	15-1 to 15-97
16	17-1 to 17-17
17	None
18	17-8
19	None
20	17-18 to 17-28

MOLECULAR BIOLOGY
AND BIOCHEMISTRY

1

A Few Basics:
Cells, Biochemistry,
Classical Genetics

Introduction

The study of molecular biology and biochemistry combines information from biology, chemistry, physics, and physical chemistry. Usually, textbooks of molecular biology and biochemistry restate fundamental material that students are expected to know from prior courses; however, biology is often considered to be so elementary that a review of biology is omitted. In this chapter, problems are provided to enable you to determine whether essential biological concepts are part of your repertoire. Necessarily, many of the problems ask only for definitions, since it is important to know the basic vocabulary of biology.

Several of the problems (1-4 through 1-7) are designed to make you think about both the smallness of cells which are used in molecular biological studies and the number of molecules contained in these cells. Some types of cells are quite small and the number of molecules per cell is likewise very small. Per cell, the number of some molecules, for example DNA, is strictly regulated and is the same for all cells of the population. However, those molecules that can freely diffuse through a cell wall or those which are themselves regulators may be present within a cell in numbers determined by the Poisson distribution. This means that individual cells in a population may at any instant be quite different from one another with respect to the concentration of a particular molecule. Averaged over a long period of time, fluctuations in the number of a particular molecule may be unimportant, but they are important if the particular molecule is an inhibitor and its concentration drops at any time below a critical value (or even to zero), for then an inhibited process may suddenly be turned on. Molecular numbers are also important when considering molecules that are synthesized on demand (for example, inducible enzymes), since a relatively small change in the number of a particular molecule per cell can result in a very large percentage change in concentration.

A large part of this chapter (problems 1-17 through 1-27) reviews simple Mendelian genetics and genetic recombination. As one studies molecular biology, it will rapidly become apparent that theories about the behavior of important biological molecules frequently grew out of simple genetic experiments. In the early period of molecular biology, these experiments merely involved either determining the relative order and positions of genes with respect to one another (that is, genetic mapping) or determining dominance relations. Problems 1-17 through 1-21 are concerned with the simplest aspect of the genetics of diploid organisms. A diploid organism containing a normal *(N)* and mutant *(n)* copy of a particular gene is said to have the genetic composition or *genotype Nn*. That is, each cell contains one copy of the *N*-type gene (called the *N allele*) and one of the *n*-type (*n* allele). When *gametogenesis* (the production of either sperm or egg) occurs and haploid germ cells are produced, the *N* and *n* alleles segregate at random; this means that in a large population of sperm cells, on the average, half of the sperm will contain the *N* allele and half the *n* allele. This will be true of the population of eggs also. If a single egg is produced, this means that there is a 50 percent chance of the egg having genotype *N* and a 50 percent chance of *n*. Thus, when a collection of sperm meets one or more eggs, since the probability of fertilization of an egg is independent of the genotype of either sperm or egg, on the average, the fertilized egg (the *zygote*) has the same probability (that is, 25 percent) of having each of the genotypes *nn, Nn, nN,* and *NN. Nn* and *nN* are the same genotypes, so this results in the familiar 1 : 2 : 1 Mendelian ratio of the progeny: 25 percent *nn,* 50 percent *Nn,* and 25 percent *NN.* An extension of this idea to a system having two or more genes is simple if one remembers that, in general, different genes assort independently. This means that if an organism has the genotype *AaBb,* the gametes *AB, Ab, aB,* and *ab* will each be produced with 25 percent probability.

These considerations enable one to determine the genotypes of the progeny. However, the *phenotype* (the observable character) depends upon whether the *N* or *n* allele is *dominant. N* is defined to be the dominant allele if the phenotype of *Nn* is the same as that of *NN.*

Problems 1-17 through 1-20 give the student practice in going through the mechanics of gametogenesis, zygote formation, and determination of the genotypes and phenotypes of the offspring.

In problem 1-21 the student is asked to perform a **pedigree analysis**—that is, to guess the genotypes of various members of a family tree if the phenotypes of some members are known. To perform such an analysis, one first writes down the genotypes of each family member whose genotype is unambiguously defined by the phenotype. For instance, if black *(B)* is dominant to blue *(b)* and a mating between a black and a blue gives one or more blue offspring, the black must have had the genotype *Bb* and not *BB* in order to contribute a blue *(b)* gamete to the blue *(bb)* offspring. Conversely, if a blue *(bb)* organism mates with an unknown parent and a black *(BB* or *Bb)* offspring results, the unknown parent must contain a *B* allele. One does not know however whether this parent was *BB* or *Bb*. Furthermore if two black *(BB* or *Bb)* parents give rise to a blue *(bb)* offspring, both parents must have had a *b* allele (each must have the genotype *Bb*).

Genetic mapping is based upon a simple principle: *when two chromosomes (or DNA molecules) interact and exchange material by genetic recombination, the probability of exchange at any one point is the same as that at any other point.* This means that the probability of an exchange occurring between two genetic markers is proportional to the distance separating these markers. The **recombination frequency,** the ratio of recombinant types to parental types, is an expression of this probability, so we may say that the recombination frequency is proportional to the distance between the two markers. This can best be seen in the following elementary example. We cross two parents with genotypes *Ab* and *aB* and 'measure the recombination frequency, *(AB + ab)/(Ab + aB)*, between *A* and *B*. Suppose this is 1 percent. This is often written: *A* × *B* is 1 percent. In another cross in which the parental genotypes were *Ac* and *aC*, we might find that *A* × *C* gives a recombination frequency of 2 percent. Thus, we may conclude that the distance between *A* and *C* is twice that of *A* and *B* (that is, 2 percent divided by 1 percent). We now know the distances but not the gene order. This can be determined by performing a third cross. Of the three possible orders *ABC*, *ACB*, and *BAC*, only one—*ACB*—is excluded by the data above since that order would require that the recombination frequency for *A* × *B* be greater than that of *A* × *C*, which is not the case. The other possibilities can be distinguished by performing the cross *B* × *C*, since for the two orders *ABC* and *BAC*, the recombination frequency for *B* × *C* would be 1 percent and 3 percent, respectively. You can gain practice in these manipulations by doing problems 1-22 through 1-27. Note that in some of the problems the gene order can be determined from a single cross; for these problems one need only remember that the probability of the occurrence of two events is the product of the probabilities of occurrence of each event. Thus, if a cross is to be made with three markers, the production of some recombinant types requires one exchange and the production of others requires two exchanges; recombinants needing two exchanges will occur with lesser probability than those needing one. Problems 1-25 and 1-27 illustrate this method.

REFERENCES

Stahl, F. W. 1964. *Mechanics of Inheritance.* Englewood Cliffs, N.J. Prentice-Hall.

Stansfield, W. D. 1969. *Genetics.* New York. McGraw-Hill. Includes 500 problems with answers.

Suzuki, D. T., and A. J. F. Griffiths. 1976. *An Introduction to Genetic Analysis.* San Francisco. W. H. Freeman and Company. Chapter 1.

Watson, J. D. 1976. *Molecular Biology of the Gene.* Third Edition. Menlo Park, Calif. W. A. Benjamin. Chapters 1, 2, 3. Superb introduction to modern biology.

Problems

○1-1. Define the following terms: Prokaryote; eukaryote; nucleolus; prophase; metaphase; anaphase; telophase; karyotype; haploid; diploid.

○1-2. Which of the following cellular components contain DNA and which contain RNA? Mitochondrion; endoplasmic reticulum; nucleus; nucleolus; chloroplast; cytoplasm.

○1-3. Describe some of the experimental evidence that chromosomes contain DNA.

○1-4. Roughly how many DNA molecules are contained in a single eukaryote chromosome?

1-5. Roughly how many DNA molecules are contained in the nucleus of a single human cell and in the cytoplasm of a single human cell? (The second number can be estimated.)

●1-6. A cell of *Escherichia coli* is cylindrical, about 1 μ in diameter and 3 μ long. At pH 7, how many H$^+$ ions are there in a volume the size of one cell? Note that, statistically, this number would vary greatly from cell to cell. Since many reactions are strongly pH-dependent, do you think that such fluctuations, if they in fact occur, would introduce significant heterogeneity among the individual cells of a population? Explain. *E. coli* is capable of normal growth in nutrient media having a wide range of pH values. This is probably because the pH within the cell is not the same as that of the surrounding medium. How do you think a cell might regulate its internal pH?

1-7. A cell of *E. coli* contains about 10^{-14} g of DNA. A DNA strand is 20 Å wide and has a mass of about 2×10^6 daltons for each μ of length. What fraction of the volume of *E. coli* is DNA?

1-8. *E. coli* has a cylindrical shape about 1 μ in diameter and 3 μ long. The doubling time of *E. coli* when growing on nutrient agar is about 25 minutes. After 12 hours of growth, a colony is roughly 2 mm in diameter and $\frac{1}{2}$ mm high. Have all of the cells been growing for 12 hours?

●1-9. Some bacteria have the property that the colonies formed on an agar surface are very large—for example, 5–10 mm in diameter. It is known that, if methyl cellulose is incorporated into the agar (this has the effect of greatly increasing the viscosity of the agar), the colonies will be much smaller. Furthermore, these colonies will frequently be surrounded by a slightly translucent film that extends over a large area. What might be the cause of this colony morphology?

1-10. What is the essential difference between an enzyme and a coenzyme? What kinds of reactions are typically carried out by coenzymes?

○1-11. Distinguish between anabolic and catabolic reactions.

○1-12. What is the principal biological role of the glycolysis reaction and the Krebs cycle?

○1-13. Which of the following are true statements?
(a) Enzymes affect the direction of a chemical reaction.
(b) Enzymes alter the speed of a chemical reaction.
(c) Enzymes are rarely, if ever, consumed in chemical reaction.
(d) Enzymes are always proteins.
(e) A particular enzyme can catalyze reactions involving many different substances.
(f) A particular enzyme frequently carries out different types of reactions involving a single chemical group.

○1-14. How is the energy generated during metabolic processes usually stored for later use?

1-15. Explain why, when glucose is the sole carbon source, bacteria grow much more slowly in the absence of oxygen than in the presence of oxygen.

1-16. Bacteria can use a very large number of compounds, such as sugars, alcohols, and amino acids as a carbon source. Animal cells usually require a single sugar—that is, they are basically glucose burners. Why might this be expected?

1-17. Hair color in some animals can be black *(BB)*, gray *(Bb)*, or white *(bb)*. If a black and gray mate and produce one offspring, what is the probability that it is gray? If there are two offspring, what is the probability that both are gray? If there are three offspring, what is the probability that only one is gray?

1-18. Two parents each with genotype *AaBb* mate. They have 16 offspring. How many would you expect to be homozygous recessive for both genes?

1-19. An animal can have red *(RR* or *Rr)* or blue *(rr)* eyes. If they are also *tt*, their eyes are colorless *(TT* and *Tt* give color). If *RrTt* mates with *RRtt*, what is the probability of getting a blue-eyed individual?

1-20. An animal has a single gene for tail shape. If a fat-tailed animal is mated to another fat-tailed one, only fat-tailed animals result. If fat and thin are mated, half are thin. If thin and thin are mated, there are always, on the average, twice as many thins as fats. Identify the genotypes of fat and of thin. *Hint:* Determine the allele that is dominant.

1-21. Color blindness in humans is inherited as a sex-linked recessive trait. Write as far as possible the genotype of each person represented in the pedigree shown in Figure 1-1. Black = colorblind.

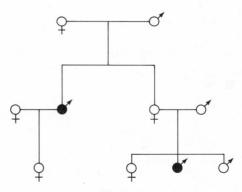

FIGURE 1-1

1-22. Following is a list of mutational changes. Which would be recessive in a heterozygote and which would be dominant?
(a) The mutant protein has no activity but the total number of molecules made by the gene is in excess of that needed for normal biological function.
(b) A protein contains 4 subunits. Both mutant and good protein units can interact. One defective subunit eliminates activity.
(c) A mutant enzyme fails to carry out a particular chemical reaction.

(d) A mutant enzyme reverses the chemical reaction of a normal enzyme.

1-23. (a) Suppose the following recombination frequencies occur between the indicated markers: $a \times c$, 2 percent; $b \times c$, 13 percent; $b \times d$, 4 percent; $a \times b$, 15 percent; $c \times d$, 17 percent; $a \times d$, 19 percent. What is the gene order?
(b) In the cross $aBd \times AbD$, what is the frequency of getting ABD?

1-24. A linear phage with genotype Abd is crossed with another with genotype aBD. The following recombinants are observed at the indicated frequencies: abd, 2 percent; ABd, 3 percent; aBd, 0.06 percent. What is the gene order?

1-25. In the cross $abcd \times ABCD$, the following recombinants were found at the indicated frequencies: $ABCd$, 3 percent; $abcD$, 3 percent; $AbcD$, 0.03 percent; $AbCd$, 0.0006 percent; $ABcD$, 0.06 percent. What is the gene order?

1-26. Three genes have the order $abde$. The recombination frequencies between various pairs are: $a \times b$, 1 percent; $b \times d$, 2 percent; $d \times e$, 3 percent.
(a) What is the frequency of production of AE recombinants in the cross $AbDe \times aBdE$?
(b) What fraction of the AE recombinants will be $ABDE$?

1-27. In a cross between two phages having genotypes EFG and efg, 1,000 progeny were analyzed. The number of phages having each of the eight possible genotypes were as follows: efg, 396; EFG, 406; eFg, 23; efG, 1; EfG, 25; Efg, 75; eFG, 73; EFg, 1. Construct a map of these markers.

1-28. An animal has a coat that is either red or white. Its tail is either long or short. When red, long is mated with red, long, only red progeny are found. When white, long is mated with white, long, both reds and whites are found. When long is mated with long, only long progeny arise. When red, short mates with another red, short, both short and long progeny arise.
(a) What are the genotypes of red, white, long, and short?
(b) If a white, short mates with another white, short, the resulting progeny have these frequencies: $\frac{1}{9}$ red, long; $\frac{2}{9}$ red, short; $\frac{2}{9}$ white, long; and $\frac{4}{9}$ white, short. Explain how these frequencies arise.

1-29. A pterodactyl can have blue or white eyes and long or short wings. Blue and long are dominant. A blue-eyed, long-winged male that is known to be heterozygous for both loci mates with a blue-eyed, long-winged female that is also heterozygous. The following frequencies of progeny are found: $\frac{3}{8}$ blue, long female; $\frac{1}{8}$ blue, short female; $\frac{3}{16}$ blue, long male; $\frac{1}{16}$ white, short male.
(a) What gene is carried on the X chromosome (assuming that sex is determined as in humans)?
An archaeopteryx can also be blue- or white-eyed and long- or short-winged. Again, a heterozygous blue-eyed, long-winged male

mates with a blue-eyed, long-winged female. This time the progeny have these frequencies: $\frac{1}{2}$ blue, long female; $\frac{1}{4}$ blue, long male; $\frac{1}{4}$ white, short male. Another heterozygous blue-eyed, long-winged female mates with the same male and there are $\frac{1}{2}$ blue, long female; $\frac{1}{4}$ blue, short male; $\frac{1}{4}$ white, long male.

(b) How does an archaeopteryx differ from a pterodactyl?

2

Early Experiments in Molecular Biology

Introduction

DNA was first isolated and partially characterized in the middle of the 19th century. However, universal acceptance that genes are molecules of DNA did not come until the early 1950's. The idea that genes might be made of protein rather than DNA had evolved in a simple and not unreasonable way as a consequence of the discovery of the chemical composition of protein molecules, and was not easily disproved. First of all, it was known that genetic material would need to be capable of enormous variations in chemical composition in order to carry the required variety of information. Proteins contain twenty amino acids, were known to be present in chromosomes, and were therefore an obvious candidate for the genetic material. On the other hand, existing data on the base composition of DNA suggested that all four bases in DNA occurred at the same frequency, so it seemed unlikely that DNA might be the genetic material even though it was found in chromosomes. The incorrect observation about the base composition was a result of two things: (1) the use of tissue of higher animals as a source of DNA, animal cells typically having equimolar quantities of the four DNA bases, and (2) inaccurate methods for separating the DNA bases for chemical analysis, which meant that small departures from equimolarity were not recognized. These data led to the proposal that DNA consisted of a repeating polymer, each unit of which contained four nucleotides in a roughly planar array; this was called the *tetranucleotide hypothesis* (see problem 2-1) and in the 1930's it provided an acceptable structure for a molecule thought to serve as scaffolding for the protein of the chromosomes.

The most critical experiments demonstrating the genetic role of DNA were those of Oswald Avery, Colin MacLeod, and Maclyn McCarty, who in 1944 discovered *bacterial transformation;* had it not been for the long-standing and tenaciously held belief that genes were made of protein, the transformation phenomenon could have provided convincing proof that DNA is the genetic material. It is worth while to read their original report, cited at problem 2-2; in summary, they showed that a bacterium having the genotype *a* can be permanently transformed to one having genotype *A* by exposure of the *a*-type bacterium to DNA extracted from an *A*-type bacterium. The principle difficulty other scientists had in accepting the idea that DNA is the genetic material arose because the DNA samples used by Avery and his co-workers were not pure but contained protein; this detail is explored in problem 2-3. Problems 2-4 and 2-7 challenge you to design a conclusive experiment that could have been done if an ultracentrifuge (which did not become standard laboratory equipment until nearly a decade after the Avery-MacLeod-McCarty experiment) had been available to these workers. Problems 2-5 and 2-6 give you the opportunity to answer other objections by designing new experiments.

Several years after the transformation experiment was reported, Erwin Chargaff capitalized on the new technique of paper chromatography, which allows efficient separation of DNA bases, and determined the base composition of DNA isolated from a large number of organisms. Two important discoveries came from this work: (1) the base composition of DNA from different organisms is not the same, yet (2) in different DNA the molar ratios guanine:cytosine and adenine:thymine both equal 1. The first point destroyed the tetranucleotide hypothesis; all objections to the work of Avery and his co-workers were dropped and the idea that DNA is the genetic material substance immediately became accepted.

The final proof for the DNA-gene idea came from the Nobel Prize-winning experiment of Alfred Hershey and Martha Chase, known as "the blendor experiment." Hershey and Chase prepared *E. coli* T2 phage samples that contained, in one case, ^{32}P-labeled DNA and, in the other, ^{35}S-labeled protein. Each of the labeled phage samples was allowed to adsorb to a sample of *E. coli* and after a brief period each of the two samples of infected cells was agitated violently in a kitchen blendor. The force of blending was capable of stripping phage particles from the bacteria; furthermore, these could be separated from the bacteria by centrifugation; with the bacteria sedimenting to the bottom of the centrifuge tube and the phage particles remaining in the supernatant. Hershey and Chase observed that after adsorption, blending, and centrifugation, ^{32}P but not ^{35}S sedimented with the bacteria; they then argued that, since bacteria from which phages had apparently been removed were still capable of phage production, in some manner the ^{32}P-containing material (the DNA) but not the ^{35}S-containing material (the protein) must carry genetic information. When the original paper, cited at problem 2-8, is read, it will be seen that *some* ^{32}P remained in the supernatant and *some* ^{35}S cosedimented with the bacteria. Therefore, the objection to the significance of the transformation experiments—that the DNA samples contained protein—could have been raised again; however, the idea that DNA is the genetic material had already become widely accepted so that this possible difficulty was not even recognized. Details of the Hershey-Chase experiment are explored in problems 2-8, 2-9, and 2-10.

REFERENCES

Freifelder, D. 1978. *The DNA Molecule: Structure and Function*. San Francisco. W. H. Freeman and Company. Section 1 presents a review of DNA research in the nineteenth and early twentieth centuries.

Stent, G., and R. Calendar. 1978. *Molecular Genetics*. San Francisco. W. H. Freeman and Company. Chapters 7 and 10. An outstanding history of the discovery of transformation and of the critical phage experiments.

Problems

2-1. An early structural model for DNA was described by the so-called tetranucleotide hypothesis: four nucleotides (one each of adenine, cytosine, guanine, and thymine) were thought to be covalently linked to form a planar unit. The units were thought to be linked together, to yield a repeating polymer of the tetranucleotide. If DNA had such a structure, why would it not have been the genetic material?

Problems 2-2 through 2-4 refer to the following paper on bacterial transformation: Avery, O., C. MacLeod, and M. McCarty, *J. Expt. Medicine* (1944), **79**:137.

2-2. Which of the following are true? The original experiments on transformation in *Pneumococcus* showed that

(a) A hereditable character of a bacterium could be permanently altered by exposure to DNA from a bacterium having a different character.

(b) A nonheritable character of a bacterium could be permanently altered by exposure to DNA from a bacterium having a different character.

(c) All genetic material must contain DNA.

(d) All of the genes in a bacterium must be carried on DNA molecules.

2-3. Following publication of the transformation experiments of Avery, MacLeod, and McCarty, opponents of the DNA-gene theory argued that the transformation was a result of protein that was contaminating the DNA sample.

(a) Suppose transformation was carried out by protein, not DNA molecules, and that the DNA preparation used contained at most 0.02 percent protein. How many protein molecules (the size of *E. coli* tryptophan synthetase—about 300 amino acids) would have been present in one milliliter of a DNA solution at a concentration of 10^{-7} mg/ml?

(b) If protein was the active agent in transformation, would the number calculated in part (a) of this problem account for the fact that, in a typical transformation experiment, 10^3 transformants result from 10^{-4} g of *Pneumococcus* DNA?

●2-4. Suppose you wish to prove to a skeptic that transformation is mediated by DNA and not by protein. You do not have available pure enzymes that degrade DNA or protein. However, you have worked hard and have shown *(i)* that 50 different genetic traits can be transformed, and *(ii)* that transformation is very inefficient and highly dependent on DNA concentration. The only piece of non-microbiological laboratory equipment you possess is a preparative ultracentrifuge and with it you are able both to measure the molecular weight of DNA and fractionate DNA according to molecular weight. Design a simple experiment to prove that the transforming principle is DNA. *Hint:* Think about linkage.

2-5. A criticism of the transformation experiments was that DNA might somehow be involved in the biosynthesis of the polysaccharide coat of the virulent *Pneumococcus*. Thus the nonvirulent mutant would be lacking some step required for polysaccharide synthesis and this step could be bypassed by the addition of DNA. By this argument, DNA would not be a genetic substance. What kinds of experiments could be done to eliminate this argument?

2-6. A critic of the interpretation of the transformation experiment might say that protein is the genetic substance and the protein can penetrate the cell only when the protein is bound to DNA. Thus the loss of transforming activity following boiling of the DNA might be a result of dissociation of protein from the DNA. Assume that you have current knowledge about the denaturation of proteins and of DNA, about the effect of low and high pH on the chemical and physical properties of DNA and protein, and about the ionic strength dependence of the binding of protein to DNA; design an experiment to prove that the critic is incorrect.

2-7. Assume *(i)* that you know how to prepare density labeled DNA as in the experiment of Meselson and Stahl, and *(ii)* that the technique of density gradient centrifugation in CsCl has been developed. Design an experiment using the techniques of denaturation and renaturation to prove that DNA, but not protein, is the genetic material.

Problems 2-8, 2-9, and 2-10 refer to the following paper: Hershey, A., and M. Chase, *J. Gen. Physiol.* (1952), **36:** 39.

2-8. (a) What experiment in the paper was used to prove that the ^{32}P was found only in the DNA and the ^{35}S was found only in the protein?
(b) What is the reasoning of the argument that phage DNA enters *E. coli?*
(●c) Why do you think that not all of the ^{35}S is stripped off by blendor action?
(●d) Why do you think that 30 percent of the ^{32}P remains in the supernatant after blending?
(e) What is the significance of the fact that phage are produced by the infected bacteria after blending?
(f) What explanation can be given for the fact that only a fraction of the injected ^{32}P is transferred to progeny DNA?
(●g) Suppose neither ^{32}P nor ^{35}S could be removed by blending; could any conclusion have been drawn? How does the result of the transfer experiment affect your answer?

●**2-9.** Suppose that when the Hershey-Chase experiment was done, the following results were obtained. After blending and centrifugation, the ^{35}S sedimented into the pellet and the ^{32}P remained in the supernatant. Microscopic observation showed that after blending, all of the bacteria had been broken into small fragments. Electron-microscopic examination of the pellet showed that there were no whole cells but just phage shells attached to small fragments of something (you don't know what). What else would you have to know, measure, or determine to enable you to draw the conclusion that DNA is the genetic material of the phage? Tell what reservations, if any, you might have about your conclusion.

2-10. You wish to prove that DNA but not protein is the genetic material. You conceive of an experiment like the blendor experiment but unfortunately all phages have short tails and cannot be removed from the cell wall with a blendor. You know how to isolate and purify progeny phages. You prepare phages labeled with both ^{32}P and ^{35}S, infect bacteria, and collect the progeny from the infection.
(a) Which of the following results could be taken as evidence to support your belief? Explain why.
(1) ^{35}S never appears in progeny phage.

(2) ^{32}P always appears in several progeny phage but the amount per phage is *much* less than the amount in the parental phage particles. ^{35}S appears in no phage particles.

(3) ^{32}P appears in only one or two progeny phage particles and in only one strand of the DNA. ^{35}S appears in many progeny phages and the amount per phage is *much* less than the amount in the parental phage particle.

(b) If DNA is the genetic material, is it possible that no ^{32}P would appear in any progeny phage? Explain your answer.

3

Protein Structure

Introduction

The sequence of amino acids in a polypeptide chain determines the three-dimensional configuration (or *conformation*) of a molecule, the dependence of conformation on various chemical and physical agents, the relative ability of the polypeptide chain to bind small molecules or other proteins and, in fact, all of the chemical and physical properties of the protein. Since proteins play a crucial role in almost all biological processes, to understand how a particular protein performs its functions is an important part of molecular biology.

In theory, if the amino acid sequence of a polypeptide chain is known (and if there are neither cross-links such as disulfide bonds nor coordination complexes produced by metal ions), the conformation of a protein can be calculated. However, except for a few relatively simple proteins, calculation of conformation is exceedingly complex and often nearly impossible. Therefore, one makes numerous physical measurements (a few of which are absorbance, circular dichroism, sedimentation, viscosity, and thermal stability) and compares the data with that obtained for proteins whose conformation is accurately known from X-ray diffraction analysis. In all determinations of conformation, even by X-ray diffraction, one usually makes educated guesses that are based upon known principles concerning the usual positions of particular amino acids. For example, it is known that certain amino acids interact with one another, that some tend to be within, and some on, the surface of a folded polypeptide chain, and that some amino acids have the potential to be components of binding sites for smaller molecules or other proteins (and will therefore be on the surface of the molecule, since it is reasonable to assume that binding sites are always on or very near the surface of the molecule).

In the hope that you, too, can someday make educated guesses about conformation, this chapter explores both the chemical and physical properties of amino acids and some of the principles of protein structure.

AMINO ACIDS

The basic structure of all of the biological amino acids except proline is

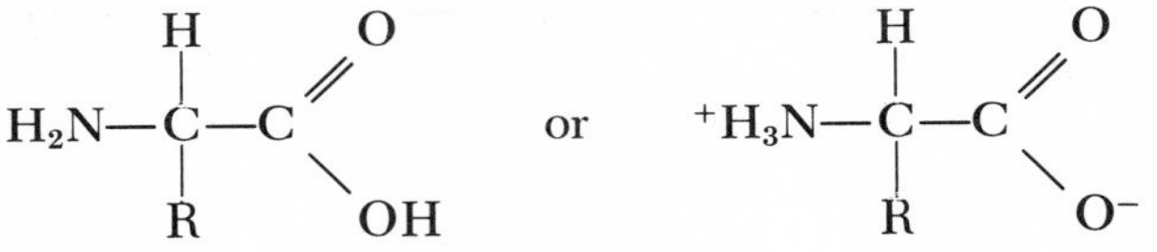

in un-ionized and ionized form, respectively, where R stands for a *side chain.* The side chain in proline is bonded to both the amino group and the α-carbon. The structures of these 20 amino acids are shown in Figure 3-1. Ionization changes with changing pH. At pH 7, both the α-amino and α-carboxyl groups (those groups attached to the carbon atom at the left in the diagrams above) are ionized (as indicated in the diagram at the right, above), and are said to be *polar.* All free amino

FIGURE 3-1   The Amino Acids and their Chemical Structures

Polar amino acids (tend to be on protein surface)

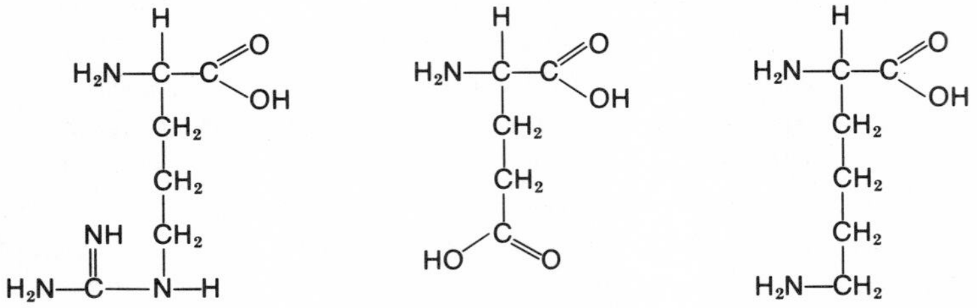

Arginine   Glutamic acid   Lysine

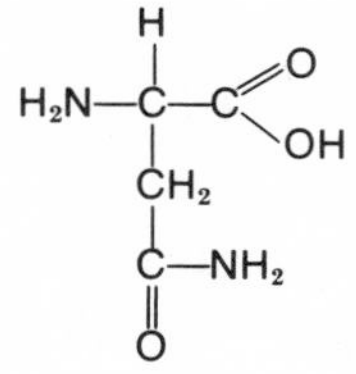

Asparagine

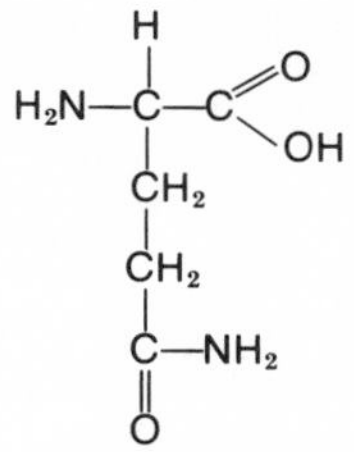

Glutamine

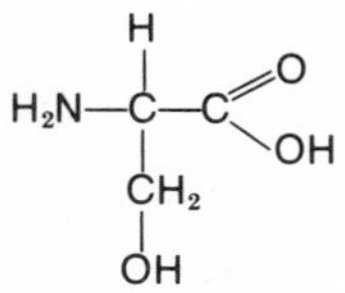

Serine

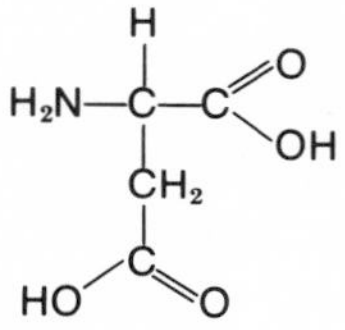

Aspartic acid

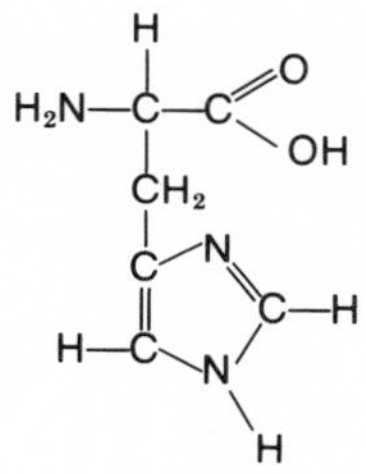

Histidine

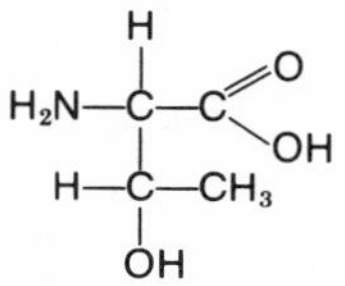

Threonine

Nonpolar amino acids (tend to be internal)

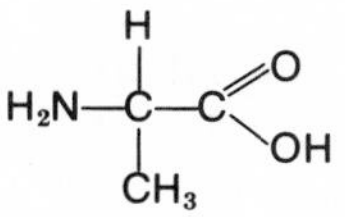

Alanine

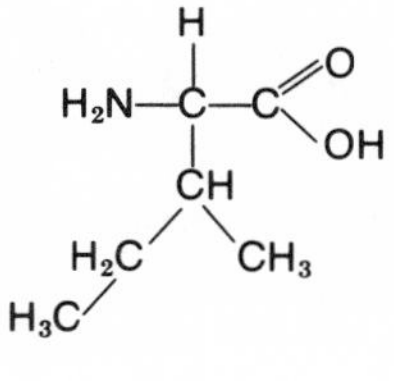

Isoleucine

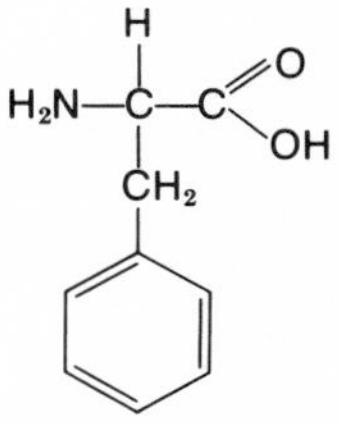

Phenylalanine

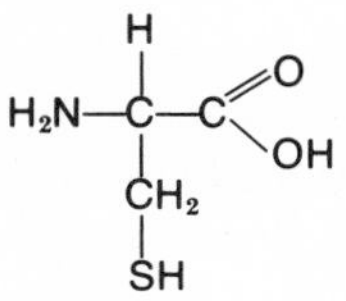

Cysteine

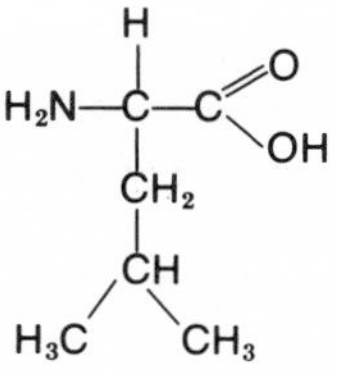

Leucine

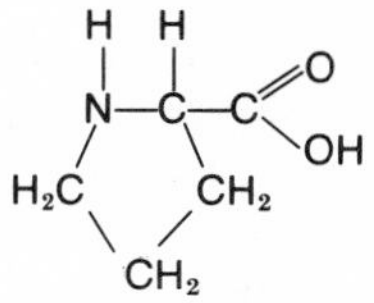

Proline

(continued overleaf)

FIGURE 3-1, *continued*

Glycine   Methionine   Valine

Amino acids equally frequently internal and external

Tryptophan   Tyrosine
(nonpolar)   (polar)

acids are polar. However, in a polypeptide chain the α-amino and α-carboxyl groups of the amino acids are engaged in peptide bonds; thus the properties of amino acids in proteins are determined primarily by the side chains. Polar side chains (those containing electronegative N, O, and S atoms) carry groups which, when charged, interact significantly with water, and some of these groups can also form hydrogen bonds to water when uncharged. Polar amino acids are therefore called **hydrophilic.** Nonpolar side chains contain only C and H atoms (except for the un-ionizable S in methionine) and cannot interact with water. Nonpolar side chains are called **hydrophobic.** We will see in the next section that the strength of the interaction of a side group with water is an important factor in determining the conformation of a protein.

Chemical properties of side chains also determine the ability of certain amino acids to bind metal ions and other small molecules. As examples, those metal ions which form insoluble sulfides or hydroxides tend to complex with amino acids containing SH or OH groups, respectively, and histidine has a lone electron pair on one of its ring nitrogens and therefore can form bonds with certain metal ions (for instance, the iron in hemoglobin is bound to histidine).

Properties of amino acids are examined in problems 3-1 through 3-7.

PROTEIN STRUCTURE

The conformation of proteins is usually determined by noncovalent interactions between amino acids. These noncovalent interactions are of several types, of which four are outlined below, in order of decreasing contribution to protein structure.

1. Hydrophobic interactions.   The side chains of nonpolar amino acids do not interact with water. Such molecules have a tendency to minimize contact with water. This is accomplished by a clustering of nonpolar groups (which reduces the surface-to-volume ratio) and a folding of the polypeptide chain so that these amino acids are usually located in internal regions of the folded chain. Hydrophobic interactions are sometimes called *hydrophobic bonds.*

2. Hydrogen bonds.   The side chains of polar amino acids interact significantly with water and in a protein chain these amino acids have a tendency to maximize contact with water and form hydrogen bonds; therefore the polypeptide chain folds in such a way that these amino acids are located on the surface of the molecule. Side chains containing amino or hydroxyl groups or a charged ring nitrogen can also form hydrogen bonds to an amino acid. Frequently the nitrogen in a peptide group forms hydrogen bonds with a peptide carbonyl group; when a series of hydrogen bonds form between peptide nitrogens and peptide carbonyl groups four residues down a chain, a structure known as an *α-helix* results.

3. Charge interactions, or ionic bonds.   Amino acids of unlike charge can attract one another, bringing different parts of the polypeptide chain together. Similarly, charge repulsion can maintain separation of different parts of a chain.

4. Minor interactions.   Other attractions between molecules are due to induced dipoles, fluctuating dipoles, and other forces. Some of these, such as van der Waals forces and London dispersion forces, are very weak but nonetheless play a role in determining protein structure. An important weak interaction is the tendency of aromatic rings to stack so that the planes of the rings are parallel. Phenylalanine provides an example of this interaction.

The three-dimensional structure of a protein is the result of an interplay between interactions that tend to draw particular amino acids together and those that tend to separate certain amino acids. For example, consider a polypeptide chain that has amino acids A and B adjacent in one part of the chain and C and D adjacent in another. Although A and C might be attracted by ring stacking, if B and D are polar and have the same charge, the attractive force between A and C will be weakened.

In a few simple cases a reasonable prediction about structure can be made if an amino acid sequence is known. For example, a peptide that consists of ten negatively charged aspartates in a row followed by three glycines and then ten positively charged lysines would probably fold to produce a hairpin configuration containing aspartates in one arm and lysine in the other. Similarly, a polypeptide consisting of twenty alternating lysines and threonines followed by six leucines would probably fold so that the lysines and threonines would surround the hydrophobic leucines. The *precise* manner of the folding is, of course, more difficult to predict.

EXTERNAL EFFECTS ON PROTEIN STRUCTURE: DENATURATION

If the attractive and repulsive forces determining the structure of a protein are varied in either quality or magnitude, the conformation of the protein will change. The somewhat vague term *denatured* is used to describe a protein if the changes are large or if biological activity is lost. The process inducing such changes is called *denaturation.* If the structure changes from one that is ordered to one lacking all order, the term *helix-coil transition* is used to describe the change. Proteins that have neither the original, *native state* nor are completely disordered are said to be *partially denatured.* Helix-coil transitions are often studied, because the parameters of the transition and the effectiveness of the denaturing agents can yield information about protein structure. Some of the commonest denaturants are discussed below.

1. pH. The pK values of ionizable groups in side chains are not all the same. The native structure of proteins is generally that which occurs in the biological pH range of 6–8. At lower or high pH values, charges in various regions of the protein can be either eliminated or created and thus attractive and repulsive interactions can change enormously.

2. Temperature. Ionic bonds between regions of unlike charge, and hydrogen bonds, have sufficiently low energy that they are easily disrupted at temperatures above about 60° C. Hence the structure of all proteins is altered by heating; the temperature required for complete denaturation depends upon the relative contribution that ionic and hydrogen bonds make to the stability of the protein.

3. Salt concentration. Charged groups frequently bind ions in solution so that a charge is neutralized. At physiological salt concentration $(0.15 M)$ most charged species that are not engaged in hydrogen bonds have bound ions. Thus at very low ionic strength (for example, 0.001 M), where many charges are not neutralized, attractive and repulsive forces that alter structure can come into play. Similarly, at very high ionic strength (for example, 1–2 M) ionic bonds may be disrupted by displacement of a charged amino acid side chain by an ion. Salt concentration can also influence hydrophobic bonds by affecting either the structure of water or the degree of avoidance of water. The interplay of forces is sufficiently complex that it is difficult to predict the precise effect of salt concentration. A rule of thumb, however, is that proteins are unstable at low salt molarity and stability increases as salt molarity increases to physiological values.

4. Chemical denaturants. The addition of certain reagents (such as urea and ethylene glycol) to solutions of proteins causes denaturation. Most denaturants disrupt protein structure by weakening hydrophobic interactions through an increase in the solubility of nonpolar side chains and by breaking hydrogen bonds. For many years the great denaturing power of urea, whose amino and carboxyl groups can form hydrogen bonds with those groups in proteins normally engaged in hydrogen bonds, was taken as evidence that denaturants acted solely by a breaking of hydrogen bonds. However, proof that a breaking of hydrogen bonds is not the mechanism by which these reagents denature came from the observation that if hydrophobic substituents are added to the denaturant, (as in *dimethyl*urea or *propyl*ene glycol) the denaturant becomes more effective (that is, the denaturant works at a lower concentration) though its ability to affect hydrogen bonds is unchanged. As the length of the nonpolar substituent increases, the solubility of nonpolar substances increases and the denaturant becomes more effective; this provided the principal evidence that hydrophobic interactions are also important in determining protein conformation.

SUBUNIT STRUCTURE OF PROTEINS

Many naturally occurring proteins are aggregates of several polypeptide chains. Each chain is called a *subunit* of the protein. Frequently subunits are identical. The interactions between these subunits are usually predominantly hydrophobic; nonpolar groups on the surface of one subunit join with those of another subunit in an attempt to avoid water. The relative orientation between subunits is determined by hydrogen bonds and ionic bonds. Subunit assembly is often spontaneous, though sometimes salt concentration must be relatively high in order to neutralize ionic repulsion.

REFERENCES

Anfinsen, C. B. 1973. "Principles that govern the folding of protein chains." *Science,* **181:** 223–230.

Dickerson, R. E., and I. Geis, 1970. *The Structure and Action of Proteins.* New York. Harper & Row. Chapter 1. The book has elegant drawings to illustrate amino acid interactions in a polypeptide chain.

Freifelder, D. 1976. *Physical Biochemistry.* San Francisco. W. H. Freeman and Company. Chapter 1.

Lehninger, A. L. 1975. *Biochemistry.* Second edition. New York. Worth Publishers, Inc. Chapters 2, 4, 6.

Stryer, L. 1975. *Biochemistry.* San Francisco. W. H. Freeman and Company. Chapter 2. Elementary and beautifully illustrated.

Watson, J. D. 1976. *Molecular Biology of the Gene.* Third edition. Menlo Park, Calif. W. A. Benjamin, Chapters 4, 6.

White, A., P. Handler, and E. L. Smith. 1973. *Principles of Biochemistry.* New York. John Wiley & Sons. Chapters 5, 7.

Wold, F. 1971. *Macromolecules: Structure and Function.* Englewood Cliffs, N.J. Prentice-Hall. Chapter 2. This is more advanced reading than the other references.

Problems

○3-1. Which amino acids are polar and which are nonpolar? Is isoleucine more nonpolar than glycine? Why? Which amino acid is unable to form a proper peptide bond?

3-2. Many proteins bind metal ions quite tightly. Which amino acids would probably be responsible for this bind in this way? If the ion were Hg^{2+}, which amino acids would most likely be involved?

○**3-3.** What is a common state of cysteine in a protein?

○**3-4.** Proteins almost always strongly absorb ultraviolet light in the wavelength range of 275–290 nm. Which amino acids are responsible for this absorption?

○**3-5.** Which amino acids can engage in hydrogen bonding?

3-6. Amino acids can be separated by electrophoresis on paper. A pH is chosen so that some are positively and some negatively charged. Thus, some will move toward the anode and some toward the cathode, the direction and rate depending on the charge. However, amino acids having the same charge, (such as alanine and valine), also separate, by moving at different rates in the same direction. Why? Which will move faster?

○**3-7.** Draw the chemical structure of the heptapeptide H_2N–alanine–tryptophan–serine–proline–leucine–isoleucine–glycine–COOH. How many peptide bonds are there?

3-8. The two least-frequent amino acids in protein X are tryptophan and methionine. Protein X contains 0.50 mole percent tryptophan and 1.50 mole percent methionine.
(a) What is the minimum number of amino acids per molecule of protein X?
(b) If protein X has a sedimentation rate that is a bit less than that of the A protein of *E. coli* tryptophan synthetase (which has 267 amino acids, what is the probable number of methionines per molecule of protein X?

3-9. A particular enzyme loses biological activity on standing in 0.01 *M* NaCl. This is prevented if the solution also contains 0.01 *M* mercaptoethanol. What information does this give you about the enzyme?

3-10. The enzyme carboxypeptidase can remove the terminal amino acid of a protein by cleaving the peptide bond formed by the terminal amino acid and the penultimate amino acid. After the terminal amino acid is removed, the penultimate one can be removed. Suppose a protein is treated with carboxypeptidase and only alanine is recovered, in the ratio of two alanines per protein molecule. What is the amino acid sequence of the terminal tripeptide?

3-11. State whether the following statements about proteins are true or false.
(a) In general, in aqueous solution, alanine is more likely to be internal than external.
(b) Serine is more likely to be internal than external.
(c) Serine is likely to be found in the active site of an enzyme.
(d) Polyalanine is more extended in 0.3 *M* NaCl than in distilled H_2O.

(e) Polyglutamic acid is more extended in $0.3\ M$ NaCl than in distilled H_2O.

(f) There is free rotation around a peptide bond.

(g) All naturally occurring proteins must contain at least one residue of each amino acid.

(h) A single amino acid change necessarily causes gross changes in protein structure.

3-12. What kinds of three-dimensional configurations might the following peptides have:

(a) H_2N–glycine–aspartic acid–methionine–alanine–alanine–glutamic acid–valine–COOH.

(b) H_2N–glycine–phenylalanine–isoleucine–glycine–aspartic acid –phenylalanine–glycine–COOH.

(c) Hexaphenylalanine.

(d) Hexaglutamic acid.

How would you expect any of these structures to vary with pH?

3-13. What would you guess to be the environment of a glutamine that is internal? What is the environment of an internal threonine?

3-14. Most enzymes lose all activity when treated with formaldehyde. Why? *Hint:* Think about the chemical reactivity of formaldehyde.

3-15. An enzyme contains 156 amino acids. (This number has no significance for the problem). Suppose amino acid 28 is glutamic acid. It is replaced by asparagine in a mutant, and all activity is lost. However, in this mutant protein, amino acid 76 is asparagine, and it is replaced by glutamic acid; full activity is restored. What can you say about amino acids 28 and 76 in the normal protein?

3-16. In aqueous solution, polymers containing both polar and nonpolar regions tend to fold so that the nonpolar regions are inside the folded structure, thus avoiding contact with water. A protein is such a polymer because some amino acids have polar and some have nonpolar side chains. Which amino acids will tend to be internal and which will tend to be on the surface of the protein? Which show little preference? Some of the polar amino acids are often found inside the protein but generally in clusters. Will cysteine also show a preference? How does cysteine differ from all other amino acids with respect to its interaction with another amino acid?

●**3-17.** Most enzymes lose activity if dissolved in distilled H_2O. Why?

●**3-18.** Most enzymes lose activity if a dilute solution is shaken violently so that foaming occurs. This can happen even if there is no decrease in molecular weight. Why is the activity lost?

3-19. A protein contains four cysteines. If all were engaged in disulfide bonds and if all possible pairs of cysteines could be joined, how many different protein structures would be possible?

3-20. Consider two proteins, each containing four identical subunits. One protein comes from rabbit, the other from dog. If either protein is heated to 57° C, the subunits separate; when the mixture is then cooled to 20° C, the subunits re-form the correct protein structure. Suppose the two proteins are mixed together, heated to 57° C and then cooled to 20° C. How many different protein types can occur if the subunits are arranged linearly? How many if the subunits are in a square array?

3-21. What kinds of forces might hold together a protein containing identical subunits?

3-22. A particular enzyme has a molecular weight of 60,000. When dissolved in $7 M$ urea, the molecular weight drops to 20,000. When the urea is removed, the molecular weight is again 60,000 and the enzymatic activity is the same as before urea treatment. A mutant enzyme shows the same changes in molecular weight but there is no enzymatic activity either before or after the urea cycle. Suppose equal weights of the normal and mutant enzymes are mixed and treated with urea, and then the urea is removed. What can you say about the requirements for enzymatic activity if the activity is now 50 percent of the starting value? 87.5 percent? 12.5 percent?

3-23. In general, partial diploids of the type z^+/z^- are Lac$^+$. Suppose you found a particular z^- mutant that produced a Lac$^-$ cell in the z^+/z^- configuration. Assuming that you know nothing about the active form of β-galactosidase, give one possible (and reasonable) explanation. The z gene codes for β-galactosidase.

●3-24. Polyglutamic acid was studied in 1:1 mixtures of (A) dioxane: $0.2 M$ NaCl or (B) dioxane: 0.01 M NaCl. Hypothetical viscosity (η) and optical rotation (α) data, as functions of pH, are shown in Figure 3-2.

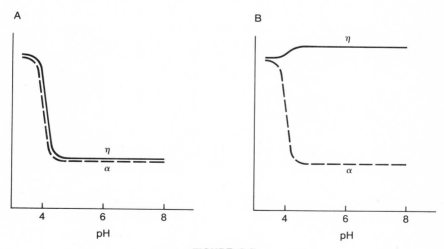

FIGURE 3-2

Explain the following:
(a) The structural alteration produced by pH change.
(b) The difference in viscosity profiles.

Predict:
(c) The viscosity and optical rotation profiles as functions of pH at high and low salt concentrations in the absence of dioxane.
(d) The viscosity and optical rotation profiles as functions of pH in 1:1 mixtures of dioxane: $2\,M$ NaCl.

4

DNA Structure

Introduction

DNA is the repository of genetic information in all organisms except for a relatively small number of viruses. As such, it must transmit its encoded information to the cell which contains it *(transcription)* and it must *replicate* in order that a progeny cell has the same properties as its parent. Furthermore, evolutionary pressures have produced a mechanism for creating new information *(mutation)* and for exchanging genes *(recombination)* so that new and better organisms can develop. These four properties, transcription, replication, mutation, and recombination, follow from the unique structure of the molecule as well as from important features of the base sequence of the molecule. Furthermore, as organisms have evolved, DNA molecules having different structures have arisen in order to carry out specialized functions often of use only to certain classes of organisms. In some cases, appreciation of the biological potential of the molecule has not preceded but followed the discovery of variations from the simple Watson-Crick model of a linear, double-stranded helix. An understanding of molecular genetics demands detailed knowledge of the various features of DNA structure. This chapter explains some of the more important features and allows you to test your ability to determine structure and to understand the relation between structure and function.

THE DOUBLE HELIX

A single polynucleotide strand of DNA is a copolymer in which deoxyribose and phosphate alternate; the side chains are single organic bases, one attached to each deoxyribose. The sugar-phosphate backbone is very polar and highly soluble in water. The bases, however, are relatively hydrophobic. Therefore, the chain tends to position the bases so that they are out of contact with water (see Chapter 3). This is accomplished in two ways: (1) the bases stack, one upon the other, thus excluding water from the surface of the heterocyclic ring, and (2) two strands interact to form a double-stranded molecule with internal bases. The hydrophobic bonding of the two strands is a major source of stability of the double helix. The familiar rules of hydrogen bonding, of adenine (A) to thymine (T) and guanine (G) to cytosine (C), also contribute to the stability and provide the specificity for the joining of the two strands.

Through ester bonds, the phosphates join the 3'- and 5'-carbon atoms of the deoxyribose. (The prime indicates that the deoxyribose is in the ring form.) At the end of the double helix, one polynucleotide strand terminates with a free 3'-OH group and the other with a free 5'-P group; for this reason the strands are said to be *antiparallel.*

STRAND SEPARATION, DENATURATION, AND RENATURATION

When heated or exposed to denaturing reagents, the two strands of the DNA double helix unwind. The optical absorbance, or *optical density,* at 260 nm (OD_{260}), increases during this process; disruption of the double helix is usually witnessed by observing this increase. A plot of OD_{260} versus either temperature or concentration of a denaturing reagent is called a *melting curve;* melting refers to the loss of the ordered structure. For thermal denaturation the temperature at which the increase in OD_{260} is 50 percent of that reached when strand separation is complete is called the *melting temperature, T_m.* AT pairs are more easily disrupted than GC pairs, so that T_m increases with higher GC content. In a melting experiment, when no further increase in OD_{260} is observed, *complete* strand separation need not have occurred, since when the last set of GC pairs breaks, this contributes only slightly to the increase in OD_{260}.

If denaturing conditions are removed (for example, if the temperature is lowered to a value at which all bases would normally be paired) before complete strand separation, the double helix will *renature* (re-form) spontaneously and practically instantaneously. This does not happen if strand separation is complete, for then a single strand will fold upon itself or join with other strands to form an aggregate having randomly arranged hydrogen-bonded base pairs. However, if aggregated DNA is heated to a temperature slightly above T_m, the random structure will melt out; given time, two strands will ultimately collide in such a way that long tracts of bases are in proper register, and complete renaturation can then occur.

SHEAR DEGRADATION

DNA molecules are very long and thin. A typical phage DNA with a molecular weight of 30×10^6 will have a length of $15\,\mu$ but a width of $0.002\,\mu$, that is, it will have an axial ratio of 7,500. If a solution of DNA is stirred or passes through a narrow tube or orifice (for example, a pipette tip), the two ends of this long strand will in general not move at the same speed; this stretches the DNA and breaks it near the middle of the molecule, if the tension is great enough. Such breakage is called *shear degradation* and accounts for the fact that when DNA is isolated from a cell containing a single DNA molecule, the molecular weight of the isolated DNA is usually much less than the weight of all the DNA contained in the cell.

Susceptibility to shear degradation increases with molecular weight and decreases with concentration since, at high concentration, entanglement of the molecules reduces the effective stretching. The shearing force increases with stirring speed and velocity of flow.

VARIANTS OF DNA STRUCTURE

Natural unbroken DNA is not usually a simple double helix consisting of two continuous strands terminating at the same site and joined at AT and GC pairs. Numerous variations exist, as listed below.

1. Single-strand breaks. When DNA is denatured, it is commonly found that the two single strands that result do not have the same molecular weight. This is caused by the presence of broken phosphodiester bonds, called single-strand breaks or *nicks*. Neither their cause nor significance is known. Sometimes one or more sugar-phosphate units are missing at the site of the break, in which case the break is called a *gap*.

2. Unusual bases. Bases other than A, G, T, and C are found in the DNA molecules of some types of phages; for example, uracil replaces thymine, and 5-hydroxymethylcytosine (HMC) replaces cytosine.

Furthermore, in DNA of the *E. coli* phages T2, T4, and T6, the HMC is glucosylated; this protects the phage DNA from phage-induced nucleases that destroy the host DNA and that would otherwise destroy the phage DNA also. Methylated adenine and cytosine are also found at low frequency in bacterial DNA; presumably, these are part of the host restriction system, which enables restriction enzymes to destroy foreign but not host DNA. Very recent experiments indicate that methylated bases may also be signals for cutting phage ϕX174 DNA in preparation for packaging in a phage coat.

3. Single-stranded termini. The DNA of the *E. coli* phage λ has a single-stranded projection 12 bases long at each end of the DNA. The base sequences are complementary so that the single strands can form a double-stranded section, thus converting the linear λ DNA to a circle. The single-stranded termini are called **cohesive sites** or **sticky ends.** A circle formed in this way is called a **Hershey circle** after its discoverer, Alfred D. Hershey. Single-stranded termini are found in DNA of a large number of different bacteriophages and may be as long as 20 bases.

4. Covalent circles and supercoiled DNA. If the two ends of a linear DNA molecule are covalently joined, a circle containing two continuous strands will result; this is called a **covalent, closed,** or **covalently closed** circle. If the two ends of the DNA molecule are rotated with respect to each other before joining, and then a covalent circle is formed, the resulting molecule will be a **twisted circle** (also called a **supercoil** or **superhelix**). For each complete rotation before end-joining, the circle will have a twist in the opposite direction. Thus, if there is one rotation, the circle will look like a figure eight. The DNA molecule can be rotated in either direction to produce molecules having either a negative or positive superhelix density (density refers to the number of twists per unit weight). The twisted circle is the usual form of viral DNA and is an intermediary in numerous biological processes.

If one or more single-strand breaks are present in a circle, the molecule is called a **nicked circle** or an **open circle.**

DETERMINATION OF THE STRUCTURE OF DNA

A variety of criteria are used to determine the structure of DNA molecules. Typically these are ways to distinguish single- from double-stranded DNA or linear from circular forms of DNA. The commonest criteria are the following:

1. Single- but not double-stranded DNA binds formaldehyde.

2. Single- but not double-stranded DNA is a substrate for *E. coli* exonuclease I, spleen phosphodiesterase, *Neurospora* endonuclease, and calf thymus polymerase.

3. The OD_{260} of double- but not single-stranded DNA increases upon heating or exposure to pH values below 2 or above 12.

4. Single- but not double-stranded DNA binds to nitrocellulose filters.

5. The sedimentation coefficient of single- but not double-stranded DNA decreases continuously as salt concentration is lowered below 0.01 molar. This is caused by repulsion of negatively charged phosphates that are not neutralized by positive counter-ions.

6. The density of single-stranded DNA in CsCl is 0.014 g/cm^3 greater than that of double-stranded DNA.

7. At neutral pH, the approximate ratio of the sedimentation coefficients of linear, circular but not supercoiled, and supercoiled DNA all having the same molecular weight is 1 : 1.14 : 1.4.

8. At pH 12.5, the approximate ratio of the sedimentation coefficients of single strands derived from linear molecules, nicked circles, and covalent circles is 1 : 1.4 : 3. The linear molecule produces two linear strands, the nicked circle produces one linear and one circular strand, and the covalent circle gives rise to two tightly entangled circles.

9. The density of covalent circles in CsCl containing ethidium bromide or propidium iodide is greater than that of nicked circles and linear molecules. The difference increases with superhelix density and is not greater than 0.04 g/cm^3.

RNA

The physical properties of RNA are for all practical purposes the same as those of single-stranded DNA (except that RNA has a higher density in CsCl). Some properties, such as circular dichroism and nuclear magnetic resonance spectra differ considerably, but discussion of these properties is beyond the scope of this book.

The chemical properties of RNA are also similar to those of DNA with the striking exception that RNA can be totally hydrolyzed to nucleotides by exposure to high pH.

REFERENCES

ELEMENTARY READINGS

Freifelder, D. 1978. *The DNA Molecule: Structure and Function.* San Francisco. W. H. Freeman and Company. This is the most recent summary of available information about the structure of DNA.

Stent, G. and R. Calendar. 1978. *Molecular Genetics.* San Francisco. W. H. Freeman and Company. Chapter 8.

Stryer, L. 1975. *Biochemistry.* San Francisco. W. H. Freeman and Company. Chapter 23.

Watson, J. D. 1976. *Molecular Biology of the Gene.* Third edition. Menlo Park, Calif. W. A. Benjamin. Chapter 9.

ADVANCED READINGS

Cantoni, G. L. and D. R. Davies. 1971. *Procedures in Nucleic Acid Research.* New York. Harper & Row.

Davidson, J. N. 1972. *The Biochemistry of the Nucleic Acids.* New York. Academic Press.

ABOUT SEQUENCING

Holley, R. W. 1966. "The nucleotide sequence of a nucleic acid." *Scientific American,* **February:** 30.

Holley, R. W., J. Apgar, G. A. Everett, J. D. Madison, M. Marquisee, S. H. Merrill, I. R. Penswick, and A. Zamir. 1965. "Structure of a ribonucleic acid." *Science,* **147:** 1462–1465. A more complete description than the above.

Mandeles, S. 1972. *Nucleic Acid Sequencing Analysis.* New York. Columbia University Press.

Maxam, A. M. and W. Gilbert. 1977. "A new method for sequencing DNA." *Proc. Nat. Acad. Sci., U.S.,* **74:** 560–564. This describes an elegant method, unlikely to be replaced, for sequencing DNA.

Problems

○4-1. What is the base sequence of the DNA strand complementary to each of the base sequences that follow?

(a) T G A T C A G G T C G A C A A

(b) T G A T C A G G A C T C A T C A

(c) A T A T A T A T A T A T A T

(d) T G A T C A G A C A G T T C T G A T T G

○4-2. Write down the names of the bases, ribonucleosides, deoxynucleosides, ribonucleotides, and deoxyribonucleotides in DNA and RNA.

○4-3. Which base pair, adenine-thymine or guanine-cytosine, contains a greater number of hydrogen bonds?

4-4. The term *nearest neighbors* in a DNA molecule refers to pairs of adjacent bases in the same strand. For a particular DNA molecule, some of the nearest-neighbor frequencies are AG, 0.15; GT, 0.03; GC, 0.08; TT, 0.10. In each case the nearest neighbor is written in the $5' \rightarrow 3'$ direction. What would be the nearest-neighbor frequencies of CT, AC, GC, and AA? If the strands of DNA were parallel instead of antiparallel, which nearest-neighbor frequencies would you know and what would they be?

○4-5. What would be the approximate molecular weight of a DNA molecule whose length is 16.4 μ?

○4-6. Order the following DNA molecules from lowest to highest melting temperature:

(a) A A G T T C T C T G A A

T T C A A G A G A C T T

(b) A G T C G T C A A T G C A G

T C A G C A G T T A C G T C

(c) G G A C C T C T C A G G

C C T G G A G A G T C C

4-7. Which of the following would have the lower temperature for strand separation? Why?

(a) A G T T G C G A C C A T G A T C T G

T C A A C G C T G G T A C T A G A C

(b) A T T G G C C C C G A A T A T C T G

T A A C C G G G G C T T A T A G A C

4-8. The following DNA molecules are denatured and then renatured. In both cases, renaturation takes place at a single temperature. Which molecule will renature with greater probability to form the original structure? Why? *Hint:* Think about intrastrand interactions.

(a) A T A T A T A T A T

T A T A T A T A T A

(b) T A G C C G A T G C

A T C G G C T A C G

4-9. The following DNA molecules are denatured and then renatured. It is found that one requires a higher temperature for renaturation. Which one is it?

(a) G A G C T G C A T C A G A T G C A G

C T C G A C G T A G T C T A C G T C

(b) A T C G G G G T A C C C C G A T A A

T A G C C C C A T G G G G C T A T T

4-10. A double-stranded hexanucleotide, for example, 3 guanines in one strand and 3 cytosines in the other, has very low thermal stability compared to a polynucleotide containing 1,000 guanines and 1,000 cytosines in each strand, respectively. Explain why.

4-11. Suppose you had a method for measuring unpaired bases in DNA and obtained the data shown in Table 4-1, at a particular temperature. How many pairs of bases would be unpaired in a polynucleotide having 100 base pairs? Where would you expect

TABLE 4-1

Number of base pairs in the DNA	Fraction unpaired (percent)
5,000	0.28
1,000	1.40
250	5.60

these unpaired bases to be? Why will they be in that position? *Hint:* See problem 4-10.

4-12. The strands of DNA separate when the DNA is placed in a solution at pH 1.5. Consider a population of DNA molecules all of which have the same molecular weight. The single strands are continuous (that is, free of single-strand breaks or nicks). After exposure of a DNA sample A to pH 1.5 for 5 minutes, $\frac{1}{3}$ (by weight) of the single strands have been fragmented. What fraction of the single strands are unbroken after 15 minutes? DNA sample B is treated in similar fashion, but $\frac{1}{3}$ or the strands are fragmented in 3 minutes. What can you say about the relative molecular weights of the DNA molecules in samples A and B? If the double-stranded DNA in A has a molecular weight of 10^7, what will be the average molecular weight of the single strands after 2 hours at pH 1.5?

○4-13. Table 4-2 shows the maximum solubilities of deoxyribose and the DNA bases in two solvents. In which solvent will a higher temperature be required to separate the strands of DNA?

TABLE 4-2

Solvent	Deoxyribose	DNA bases
A	4.3 molar	0.02 molar
B	0.5 molar	0.22 molar

○4-14. When a DNA molecule is boiled, the strands come apart. Which of the following are possible explanations for this?

(*Do not concern yourself with whether the following statements are true—only with whether they could be explanations.*)

(a) The energy of thermal vibrations is greater than the energy of the weak bonds stabilizing the double-stranded structure.

(b) The solubility of adenine and guanine is greatly increased at high temperatures and the solubility of the deoxyribose is unchanged.

(c) The solubility of adenine and guanine is greatly decreased at high temperature and the solubility of the deoxyribose is unchanged.

(d) The solubility of the bases is unchanged but the solubility of the deoxyribose is increased.

●4-15. When DNA is placed in distilled water, the two strands come apart. Explain why. *Hint:* Consider the effect of counter-ions on the interaction between charged groups.

4-16. Consider a protein that is very polar (that is, highly charged) but has a small, very nonpolar region. The nonpolar region binds very

tightly to the bases in DNA. The addition of this protein will cause
(a) The DNA to precipitate.
(b) The two strands to come apart.
(c) The DNA to become more stable.

4-17. When DNA is heated to 100° C, the two strands separate. When cooled again, in general the two strands do not come back together again because they do not collide with one another in such a way that complementary base sequences can form. In other words, heated DNA remains as single strands. Consider a DNA with structure

A T A T A T A T A T

T A T A T A T A T A

If it is heated to 100° C and then cooled, what might be the structure of the single strands? Assume that the concentration is so low that the two strands never find one another.

●4-18. Suppose you have a DNA molecule with a random sequence of base pairs and you are able to obtain a double-stranded RNA molecule having exactly the same sequence of base pairs. If these molecules are each dissolved in the same solvent (for example, dilute NaCl), which will have the greater thermal stability—that is, which will have to be heated to a higher temperature to separate the strands? Why?

●4-19. If DNA having an optical density at 260 nm (OD_{260}) equal to 1.00 is boiled, OD_{260} increases to 1.37. This is an indication of the conversion of double-stranded to single-stranded DNA. When returned to 25° C, OD_{260} remains the same, if the DNA is in a solution of low ionic strength and contains no polyvalent cations (for example, 0.01 M NaCl), but drops to 1.12 if the ionic strength is high (for example, 0.5 M NaCl).
(a) Explain the above.
(b) By various chemical means (such as by treatment with nitrous acid) the two bases in a base pair can be covalently joined. Predict the changes in OD_{260} when this DNA is boiled and then returned to 25° C. Assume the DNA is in 0.5 M NaCl.
(c) Predict OD_{260} changes for a double-stranded deoxynucleotide containing G in one strand and C in the other; this is abbreviated as poly(dG:dC). Assume the ionic strength is high.
(d) Predict OD_{260} changes for poly(dG:dC) if the polymer is circular. Assume the ionic strength is high.
(e) A DNA molecule whose structure is

A T G C A T A T A T G C A T

T A C G T A T A T A C G T A

is boiled in $0.5\,M$ NaCl and then returned to $25°$ C. What will be the ratio of OD_{260} before boiling and after boiling and cooling?

•4-20. When a DNA solution is heated and OD_{260} is measured as a function of temperature, a plot of OD_{260} versus temperature is called a melting curve. A typical plot is shown below. In such curves the indicated temperature is always that at which OD was measured. Another kind of melting curve can be obtained by heating the DNA to a temperature T, cooling the DNA to $25°$ C, *then* measuring the OD_{260} (at $25°$ C) and plotting this value of OD_{260} versus T. Such a melting curve is called an ***irreversibility curve*** or *i*-curve since an increase in OD_{260} occurs only if DNA molecules remain single-stranded. For the melting curve shown, draw very carefully a hypothetical *i*-curve for randomly fragmented bacterial DNA, intact phage DNA, and phage DNA containing several phosphodiester breaks in the single strands.

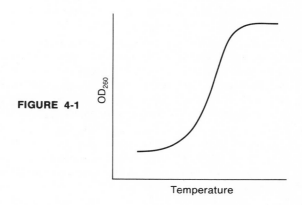

FIGURE 4-1

4-21. Bacterial DNA can be density-labeled if the bacteria are grown in a medium containing a heavy isotope (such as ^{15}N, or ^{13}C). If both strands are so labeled, the DNA is said to be heavy (HH) as opposed to normally light (LL) DNA. If equal amounts of HH and LL DNA are mixed, heated to $100°$, and slowly cooled to allow renaturation to occur, 25 percent of the DNA will be HH, 25 percent will be LL and 50 percent will be HL (hybrid). Suppose $45\,\mu g$ HH DNA and $5\,\mu g$ LL DNA are mixed, heated to $100°$, and renatured, how many μg of HH, HL, and LL will result?

•4-22. Double-stranded DNA is converted to single-stranded DNA at high pH (11.3–11.6, depending on the DNA). If the DNA contains 5-bromouracil (B), the critical pH is about 10.5. This conversion can be detected by centrifugation in CsCl since the DNA increases in density by 0.014 g/cm³ when converted to single strands. Normal DNA and B-DNA can also be distinguished in CsCl since the density of the latter is about 0.1 g/cm³ greater than the former.

At one time it was claimed that *E. coli* DNA is four-stranded, and that it consists of two double helixes held together by "biunial" bonds. Thus, the DNA of hybrid density observed by Meselson and Stahl was believed to consist of one LL duplex bonded to one HH duplex. Using growth for one generation in medium containing 5-bromouracil in order to obtain hybrid DNA, Robert Baldwin and Eric Shooter were able to distinguish BL from BB:LL DNA (that is, four-stranded DNA containing BB and BL).

Draw the curves for density versus pH expected for BL and BB:LL DNA.

4-23. If DNA is put into 100 percent methanol, its optical density increases by 37 percent. If the methanol concentration is reduced 20-fold by dilution of the methanolic solution with water, normal optical density and a double-stranded structure is restored. Explain.

●**4-24.** At pH 12, strands of DNA separate because hydrogen bonds break. What is the cause of the breaking of hydrogen bonds? Why are the hydrophobic forces unable to maintain the double-stranded structure?

●**4-25.** It has been stated that a single-stranded polynucleotide is helical if bases are stacked and this stacking tendency has been used to explain the stability of DNA. Since stacking in a single-strand polynucleotide is between adjacent bases, how can stacking explain the tendency of the two strands of the double helix to stay together? The principle here is the same as that in problem 4-24.

●**4-26.** When heating double-stranded DNA, the maximum OD_{260} reached is usually 37 percent higher than the OD_{260} at 25° C. However, in a solution of 6 M sodium trifluoroacetate, the maximum OD_{260} reached is only 16 percent higher than that at 25° C. Explain why this is so.

●**4-27.** Suppose you have obtained a melting curve but the total increase in optical density is only 20 percent, the transition being normally sharp. How might you explain this? Would T_m from such a measurement be believable?

4-28. Suppose that you have just prepared two DNA samples, one native and one denatured, and have dialyzed each against 0.01 M NaCl. You do not know the OD_{260} values before dialysis. You then add a very small amount of an enzyme to each and in so doing mix up the samples. How can you determine the identity of each sample by an absorbance measurement? You may assume that a small part of each sample is consumed in the testing and that the enzyme will not interfere with the test. Design a second test that does not require the introduction of an agent or condition that causes denaturation.

4-29. Why does salt concentration affect T_m? *Hint:* See problem 4-15.

4-30. Would you expect melting curves to be affected by the molecular weight of a DNA molecule?

○4-31. Describe several ways in which a linear DNA molecule can form a circle.

4-32. A supercoil can be made by which one of the following operations.
(a) Joining the ends of a linear DNA and doing nothing else.
(b) Twisting the two ends of a linear DNA and then joining the ends together.
(c) Joining the ends of a linear DNA and then twisting the circle.

4-33. DNA is easily broken by hydrodynamic shear forces. These forces act in such a way that a DNA molecule breaks near its middle. By slowly increasing the magnitude of the force a molecule can be successively broken into halves, then quarters, then eighths, and so on. Let us call the values of a force just sufficient to break a *linear* molecule of mass M into halves $F_{1/2}$; for quarters and eighths we use the higher values $F_{1/4}$, $F_{1/8}$, and so on. Suppose you have a *circular* molecule of mass M. Approximately what force is needed to break the molecule? What is the size of the fragments?

4-34. A Hershey circle is formed by heating a linear DNA molecule that has single-stranded cohesive ends to a temperature that favors joining of the ends. In solutions with very low DNA concentrations, *only* Hershey circles are formed. What other structures might be expected in solutions with high DNA concentrations?

The next seven questions (4-35 through 4-41) refer to the fact that super-coiled DNA molecules can be distinguished from linear and open circular forms by adding ethidium bromide to a CsCl solution containing various DNA molecules and centrifuging to equilibrium. The ethidium bromide molecules intercalate between the DNA base pairs.

○4-35. Why does a supercoil have a higher density than a linear molecule when ethidium bromide is present?

○4-36. Would you expect the density decrease induced by the presence of ethidium bromide to be affected by GC content?

4-37. In CsCl containing ethidium bromide, a supercoiled and a linear DNA have densities 1.592 and 1.556 g/cm³, respectively. What would be the density of a supercoil linked, as in a chain, with an open circle of (a) identical mass or (b) twice the mass? (Linked circles do exist and are called *catenanes*.)

●4-38. Might there be a density shift due to binding of a nonintercalating substance to a phosphate group or the deoxyribose? Under what circumstances can the nonintercalating substance be used to separate supercoiled from linear DNA? (Thinking about problem 4-39 will help to answer this problem.)

4-39. An endonuclease known as S1 makes a strand break only with single and not with double-stranded DNA. However, it can cleave one of the forms of polyoma DNA, usually making only a single break. Which form is cleaved and why?

4-40. A sample of DNA gives one sedimentation band in CsCl but two in CsCl containing ethidium bromide. The ratio of the areas of the denser to the lighter band is $2:1$. The molecular weight of the DNA is 30×10^6. Suppose that the DNA is treated with an enzyme that produces, on the average, one single-strand break in each molecule. What would be the ratio of the band areas after such a treatment? (Remember to use the Poisson distribution to determine the fraction of molecules that are unbroken.)

●**4-41.** The sedimentation-velocity properties of a supercoiled DNA molecule are being studied as a function of the concentration of added ethidium bromide. It is found that s decreases, reaches a minimum, and then increases. Explain. *Hint: s* is a function of both molecular weight and shape.

●**4-42.** You are given a homogeneous nucleic acid sample with the following properties. What is your conclusion from each bit of information and what is your final conclusion concerning the structure of the nucleic acid?
(a) It reacts with formaldehyde.
(b) Upon denaturation its buoyant density in CsCl increases by 0.010 g/cm^3.
(c) The thermal denaturation profile gives a biphasic increase in OD_{260}.
(d) The DNA is not susceptible to either *E. coli* exonuclease I or spleen phosphodiesterase, even after alkaline phosphatase treatment.
(e) It acts as a template for calf thymus DNA polymerase.
(f) Its sedimentation coefficient is somewhat higher in high ionic strength than low ionic strength.
(g) It is retained by a nitrocellulose filter.
(h) At low ionic strength, the sedimentation coefficient overestimates its molecular weight.
(i) Band centrifugation of the denatured material gives rise to two bands, A and B. Exposure of purified B for a short period of time to pancreatic DNase results in a molecule C with a single break. The sedimentation coefficient of C is about the same as B at high ionic strength, but lower than B at low ionic strength. The sedimentation coefficient of C is higher than A at low and high ionic strength.
(j) Assaying the biological activity after ^{32}P suicide indicates that about one disintegration in seven is lethal. For double-stranded phage nucleic acids, one disintegration in ten is lethal.

●**4-43.** A molecule A of a phage nucleic acid has a structure compatible with the properties listed below. What is your conclusion from each property listed? What is the structure of A, B, and C?
(a) A does not react with formaldehyde; it passes through a nitrocellulose membrane filter; and it does not act as a template for calf thymus DNA polymerase.

(b) During centrifugation, under alkaline conditions, three sedimenting species are observed.

(c) Exposure of A to E. coli exonuclease III, after which A is left to stand in 7.2 M perchlorate at 25° C for several hours, gives rise to species B.

(d) Exposure of B to DNA ligase results in partial conversion to C.

(e) C sediments faster than A under alkaline conditions and also has a lower buoyant density than A in CsCl containing ethidium bromide.

(f) Denaturation of A, followed by its interaction with polyG, gives three bands in CsCl.

(g) Exposure of each molecule of A to polynucleotide kinase in the presence of γ-^{32}P-labeled adenosine triphosphate results in the ^{32}P labeling of three nucleotides.

(h) The ^{32}P-labeling of the nucleotides in (g) takes place only if A is pretreated with E. coli alkaline phosphatase.

(i) Exposure of A to acridine dyes results in an increased viscosity and a decreased sedimentation coefficient.

(j) When A is sheared to half-molecules, there result two close bands in a CsCl density gradient, but only one band is found in a Cs_2SO_4 density gradient.

4-44. Use the following clues to determine the structure of a phage nucleic acid. What is your conclusion from each clue and what are the structures of A, B, C, and D?

(a) Molecule A, when isolated from the phage, has a length of 20 μ. At room temperature and when heated to 60° C in 0.2 M NaCl, it passes through a nitrocellulose membrane filter.

(b) Molecule A can serve as a primer for highly purified E. coli DNA polymerase. When each of the nucleotide triphosphates is presented individually to the enzyme, only guanine is incorporated.

(c) E. coli exonuclease III converts A to B after brief incubation. A and B can interact to form C when heated to 10° C below the T_m and then cooled slowly. The product C, after being treated with ligase, gives rise to two sedimenting species in an alkaline gradient; the molecular weight of the larger material is twice that of the smaller. Exposure of A to the heating and cooling cycle does not alter its size or shape.

(d) The phage that contains A can lysogenize E. coli. When the lysogenized cells are induced, a structure D can be isolated from the cells, which sediments faster in an alkaline sucrose gradient than the product of ligase action on C.

(e) Several enzymatic steps are required to convert D to A. What do you propose these steps might be?

(f) When A and D are centrifuged in ethidium bromide–CsCl gradients, they can be separated from one another. If the ethidium bromide is not present in the CsCl gradient, A and D cannot be resolved.

(g) Denatured A, when exposed to polyIG, gives two bands, after preparative centrifugation, in a CsCl density gradient; the two bands can be separated and the molecular structures in each can be isolated. Phage-specific RNA isolated from infected cells, in which phage-DNA replication has been inhibited, hybridizes only with the DNA of the two bands.

4-45. A phage mutant has been isolated whose DNA has the following properties. What is its structure based on the following properties? Discuss each property in several sentences.

(a) DNA, isolated from the phage, sediments as a single component (structure A) with an *s*-value relatively independent of the ionic strength.

(b) A does not react with formaldehyde. Acridine dyes cause viscosity to increase and the sedimentation coefficient to decrease.

(c) Sedimentation at pH 12.5 reveals two sedimenting components; there is twice as much material (measured by optical density) sedimenting in the faster than the slower component.

(d) When A is heated to 10° *below* the T_m and cooled rapidly, two components are produced, B and C, which have different sedimentation coefficients. When cooled slowly, only one sedimenting component is seen; it has an $s_{20,w}$ value the same as A.

(e) Limited exposure of A to exonuclease III, followed by annealing, results in a structure D that has a higher sedimentation coefficient than A.

(f) Treatment of D with ligase produces E, which, in alkali, sediments as a single component at least four times faster than the larger, denatured component of A.

(g) When polyG is added to heat denatured A, two bands differing in buoyant density are seen in CsCl density gradients. After the polyG is removed, phage-specific mRNA ("early" and "late") hybridizes exclusively with the DNA in the denser band of the CsCl gradient.

(h) If A is treated with DNA ligase, then terminally labeled with ^{32}P by using polynucleotide kinase, then digested to mononucleotides, two nucleotides, dAMP (deoxyadenosine monophosphate) and dGMP (deoxyguanosine monophosphate), are ^{32}P-labeled. If after the labeling but before digestion, the strands are fractionated as in (g) above, then the labeled dGMP can be shown to be associated with the strand that binds more polyG.

4-46. A nucleic acid molecule containing 50 percent GC pairs has the properties listed below. What structural feature is indicated by each property? Draw the structures of the parent molecule and of the forms A, B, C, and D that can be derived from it by accepted techniques.

(a) The denaturation temperature and buoyant density in CsCl is higher than that of *E. coli* DNA.

(b) It is insensitive to RNase.

(c) A single scission by DNase gives rise to A, with a lower s value.

(d) Denatured A (which is called B) gives rise to two bands when centrifuged to equilibrium in neutral CsCl, but only one in alkaline CsCl.

(e) Exposure to pH 13 for 60 minutes at 37° C and reneutralization gives rise to a structure (C) which has a lower buoyant density in neutral CsCl than before alkaline treatment.

(f) C is resistant to snake venom phosphodiesterase.

(g) When C plus the pH 13 digestion products are filtered through a nitrocellulose membrane filter, half of the material passes through the filter; this is called D. The material retained on the filter can be eluted and is found to be C.

(h) C interacts with polyG at 0° C, forming a complex having a buoyant density in CsCl greater than that of C.

(i) Highly purified DNA polymerase preparations can use the nucleic acid as a template in the presence of suitable deoxyribonucleotide triphosphates. When the product of two or more rounds of copying are examined for nearest-neighbor frequencies, ApC is not equal to GpT.

(j) D inhibits the action of *E. coli* endonuclease I.

(k) Polynucleotide kinase does not phosphorylate either C or D in the presence of ATP.

●4-47. A new DNA species X has been isolated from antibody-forming cells that have been treated with cycloheximide. It is not mitochondrial and has a unique density in CsCl which distinguishes it from other cellular DNA species. Based on the following properties, what is the structure of X? State your conclusion from *each* clue given and give structures for the intermediates A, B, and C.

(a) The molecular weight M determined by light scattering is 10^7, yet if the standard s-M relation valid for linear double-stranded DNA is used, the s-value at neutral pH indicates a higher M.

(b) Under neutral conditions there is only one sedimenting species of X; under alkaline conditions there are two species, A and B, and the sedimentation coefficient of B is about four times that of A. The amount of A is the same as the amount of B.

(c) When A is separated from B by preparative centrifugation in alkaline sucrose gradients, collected, and neutralized, then subsequent renaturation of A (now called A_1) makes it insensitive to hydrolysis by the *Neurospora* endonuclease. Before renaturation it is retained by filtration through a nitrocellulose membrane filter.

(d) The buoyant density of B is greater than that of A in CsCl–ethidium bromide gradients.

(e) When B is collected and reneutralized but is not deliberately exposed to renaturing conditions, it is insensitive to the *Neurospora* endonuclease and passes through a nitrocellulose membrane filter.

(f) When B is treated with the restriction enzyme of *Hemophilus influenzae,* it produces C, which has a lower sedimentation coefficient than B but the same molecular weight determined by light scattering.

(g) The $C_0 t$ value obtained in experiments on the renaturation of sheared, denatured A is the same as that obtained for sheared, denatured X DNA (at the same total DNA concentration).

●4-48. A new viral nucleic-acid structure, A, is isolated. The following clues are intended to give some information concerning its structure. Write out, in a sentence or two, the conclusion you arrive at from each clue. Draw a structure for A and for *each* of its conversion products.

(a) When A is passed through a nitrocellulose membrane, it is *not* retained on the filter.

(b) Sedimentation of A in organic denaturing solvents gives rise to a single species accounting for all of the OD_{260} of the sample; however, when A is sedimented in an alkaline gradient, it is converted to B, whose molecular weight is lower, and some OD_{260} is measurable at the meniscus.

(c) When B is filtered as in (a), it is retained on the nitrocellulose filter. When B is heated and cooled rapidly, there is no increase in OD_{260} over that of unheated B. The latter is also true for A.

(d) A is not susceptible to exonucleases, even at elevated temperatures.

(e) A is a poor template for any of the three *E. coli* DNA polymerases in the absence of a primer.

(f) When a single nick is introduced by mild treatment with pancreatic DNase, A is converted to structure C. When C is heated and then cooled rapidly, there is no increase in OD_{260} over that of unheated C.

(g) Structure A can be partially digested by ribonuclease III and nicked by either the S1 or the *Neurospora* endonucleases.

(h) B can be potentially converted to A if exposed first to DNA polymerase I and then to DNA ligase. What treatment must B undergo before polymerase I and ligase can successfully convert it to A?

(i) When cleaved by the restriction enzyme EcoR1, A is converted to D, E, and F. F has a higher buoyant density than D or E in CsCl or Cs_2SO_4 density gradients. When either D or E is heated and cooled rapidly, there is no increase in OD_{260} beyond that of the unheated samples.

●4-49. Many of the observations about phage DNA with cohesive sites at the ends might be explained by an alternative model for cohesive sites diagrammed below, where c' denotes bases complementary to c, b' to b, and a' to a.

a b c _____ c'b'a'

Assume the single-stranded regions are 12 bases long, as for λ DNA. Propose one physical and one enzymatic experiment which, if done, would distinguish this structure from λ-style cohesive ends. Your proposal should not require visualization with the electron microscope of a double-stranded piece 12 bases in length, because that cannot be done.

●4-50. DNA extracted from λ phage is linear and double-stranded, but carries 5'-P, single-stranded, complementary ends (cohesive or "sticky" ends) that are 12 nucleotide residues long. Suppose that λ DNA is:

(1) treated with the enzyme bacterial alkaline phosphatase;
(2) treated with γ-^{32}P-ATP (adenosine triphosphate) and polynucleotide kinase;
(3) allowed to stand for a week at 5° C in 0.1 M NaCl at low DNA concentration in order to allow the formation of Hershey circles;
(4) treated with DNA ligase; and finally,
(5) digested with spleen phosphodiesterase and DNase II (micrococcal nuclease).

The products of digestion are chromatographically analyzed. The only radioactive product is ^{32}P-GMP (guanosine 3'-monophosphate).

What information can be derived from these findings?

4-51. Human adenovirus causes upper respiratory infections. This virus contains double-stranded DNA of 20×10^6 daltons. When light and heavy strands of the DNA are separated and observed with the electron microscope, the structure shown in Figure 4-2 is seen. If the DNA single strands are pretreated with exonuclease III, then these circular forms are not seen. What can be said about adenovirus DNA structure?

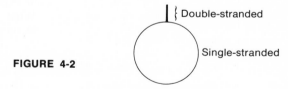

FIGURE 4-2

Adenovirus DNA structure

●4-52. A unique replicating viral nucleic acid A has the following properties. In a sentence or two state what you conclude from each clue listed below and draw a figure for each of the structures A through F.

(a) When purified preparations of A are thermally denatured and OD_{260} is followed as a function of temperature, a biphasic thermal denaturation curve is observed; initially there is a relatively rapid rise (accounting for about 8–10 percent of the total hyperchromicity), followed by a plateau and then a sharper OD_{260} increase at higher temperatures.

(b) When A is passed through a nitrocellulose membrane filter, it is retained.

(c) Preparations of A can act as a substrate for polynucleotide kinase as well as stimulate DNA synthesis by DNA polymerases I, II, and III.

(d) Native preparations of A sediment as one species; when A is denatured by heat or alkali, two sedimenting species B and C are seen. When C, which is larger than B, is treated very briefly with DNase, both it and its digestion product sediment as two components at low ionic strength but as one component in high ionic strength.

(e) Brief treatment of A with a nonspecific exonuclease converts it to D. Treatment of D with polymerase I, ligase, and appropriate additives converts D to E.

(f) When a mixture of E and A are centrifuged in an ethidium bromide–CsCl gradient, they can be separated from one another. (Which has the higher buoyant density?)

(g) Structure E, when treated with *Neurospora* endonuclease, which is specific for single-stranded DNA, is converted to F. Both A and F have similar densities in ethidium bromide–CsCl density gradients.

(h) The molecular weight of A determined in CsCl gradients is 10×10^6; by electron microscopy it is 8×10^6; and by C_0t analysis it is 4×10^6.

4-53. A linear phage-DNA molecule is hybridized with a molecule that is identical except that the central 10 percent of the molecule is deleted.

(a) What is the structure of the heteroduplex that would be seen by electron microscopy?

(b) If the missing 10 percent is replaced by a piece of nonhomologous bacterial DNA having the *same* length, what will the structure of the heteroduplex be? How will this differ if the nonhomologous bacterial DNA is only half the length of the deleted phage DNA?

4-54. Phage mu contains double-stranded DNA and infects *E. coli*. Mu is a "mutator" phage which causes mutations by inserting its genome into *E. coli* genes at random locations. When a strain of *E. coli* lysogenic for mu is grown and induced, many phage are released. All the phage particles make plaques. If the phage DNA is extracted, denatured, and renatured, the heteroduplex structures shown in Figure 4-3 predominate. (Heavy and light lines represent double- and single-stranded DNA, respectively.)

(a) Is the mu genome circularly permuted?

(b) Is it terminally redundant?

(c) What is the difference between the left and right ends of mu DNA?

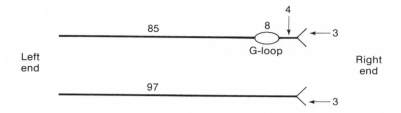

Numbers = percent of length of entire DNA molecule

FIGURE 4-3

(d) The G-loop is a region accounting for 8 percent of the mu DNA, and it is known to represent an inversion* of the DNA in that region. That is, during growth of the phage after induction, the DNA in the G-loop region becomes inverted in some phage DNA molecules. Because of this, another rare structure is seen in the population of renatured DNA molecules. Draw this structure.

(e) When an F′ factor carrying the mu prophage is mixed with DNA extracted from the phage particle, denatured, and allowed to renature, the structures shown in Figure 4-4 are seen. What do the structures tell you about the relation of the mu vegetative map to the mu prophage map? *Hint:* Think about λ phage and λ prophage and the heteroduplex map that could be observed for λ in a similar situation.

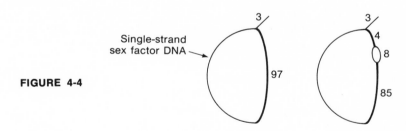

FIGURE 4-4

Numbers = percent of length of original molecule

. (f) Sheared *E. coli* DNA is found to hybridize with 3 percent of the sequences in mu DNA. The hybridization occurs equally well with all regions of the *E. coli* chromosome.

    (1). Do you expect mu to be a transducing phage?

    (2). What part of the mu DNA is hybridizing with *E. coli* DNA?

---

* This means that the DNA in a G-loop region can exist in either of two possible opposite orientations (that is, 1234 or 1324).

4-55. Figure 4-5 shows a C_0t curve for an imaginary eukaryote.
(a) What fraction of the DNA is contained in components that are unique? What fraction is redundant? What fraction is satellite?
(b) If the unique sequences are represented once per genome, how many copies per genome are there of each of the other two classes?

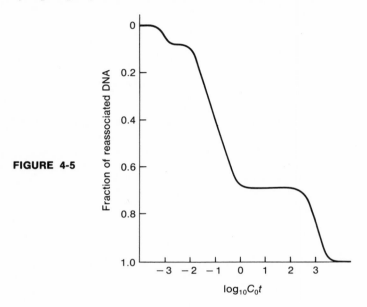

FIGURE 4-5

4-56. Suppose you are characterizing three DNA-binding proteins A, B, and C, by studying the effect of the proteins on the sedimentation coefficient (s) of superhelical DNA; you have obtained the results that are shown in Figure 4-6.

FIGURE 4-6

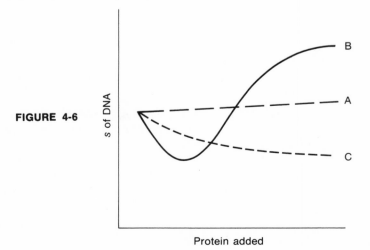

What can you say about the binding properties of proteins A, B, and C?

●4-57. Read Cozzarelli, N., R. Kelly, and A. Kornberg, *Proc. Nat. Acad. Sci., U.S.* (1968), **60**: 992.

(a) How was the covalently closed plasmid DNA separated from circular duplexes with single-strand breaks and from linear forms? Why does this method work?

(b) Explain the kinetics seen in the DNase results shown in Figure 1 of the Cozzarelli-Kelly-Kornberg paper. Why does the kinetics "strongly support a closed circular duplex structure"?

(c) Why does "the presence of tertiary turns in a DNA molecule increase its sedimentation coefficient"? Why is this effect especially marked in alkali?

(d) Two obvious possibilities concerning how this DNA molecule replicates are: (1) it is a piece of DNA, created by illegitimate recombination, from the region where *E. coli* replication has its origin, and thus parasitizes the complete *E. coli* replication system, or (2) it is a minimum viroid which codes for its own "initiator protein" (for example, like phage ϕX174). Design convincing and reasonable experiments to distinguish these two possibilities from one another.

(e) If this plasmid makes no protein, what predictions might you make, based on the Helmstetter-Cooper model for initiation control, about (1) the levels of initiator protein in cells carrying the plasmid and (2) the timing of initiation of plasmid DNA synthesis in cells of different growth rates?

●4-58. Assume that an F′ factor that already has an IS2 sequence has a second IS2 sequence inserted, but as an inverted sequence. Single-stranded F′ DNA, containing both IS sequences, is examined by electron microscopy. What type of structure would you expect to observe? (See Saedler, H., H. J. Reif, S. Hu, and N. Davidson, *Mol. Gen. Genetics,* (1974), **132**: 265–289.)

●4-59. In general, the viscosity (η) of a solution of a macromolecule increases with both molecular weight and axial ratio (ratio of length to width). The viscosity of a DNA solution decreases with time if DNase is added, owing to the production of double-strand breaks in the DNA. Single-strand breaks have no effect on viscosity. Assume that single-strand breaks are made linearly with time during exposure to an enzyme; draw a plot of the expected change of viscosity with time for a double-stranded DNA molecule which is (a) linear, or (b) circular.

●4-60. Double-stranded DNA can be converted into single-stranded DNA either by heating in the presence of formaldehyde or by adjusting the pH to 12.3. The sedimentation patterns of DNA denatured by either procedure are normally identical. However, if DNA is mixed with acridine orange and then irradiated with low doses of light at wavelengths absorbed by acridine orange, then the two procedures give different sedimentation patterns. That is, with the heat-formaldehyde treatment, the sedimentation pattern of a DNA molecule is the same as that of unirradiated DNA—a single, sharp boundary with $s_{20,w}$ roughly 30 percent greater than that of native DNA; by the alkaline treatment, there is still a small, sharp boundary, but a great deal of material sediments more slowly. Explain this difference.

●4-61. The folded *E. coli* genome has a high s-value and low viscosity. Why is this so? How can the value be lowered and the viscosity raised? How would you identify the physical and genetic region of the *E. coli* genome that is associated with the cell membrane? How could one show whether a particular membrane-associated protein is bound specifically to the *E. coli* genome?

●4-62. Explain why RNA is hydrolyzed by alkali, whereas DNA is not.

4-63. How would you distinguish, by several methods, single-stranded DNA from single-stranded RNA of the same molecular weight?

●4-64. How would you distinguish 23s ribosomal RNA from phage RNA of the same size?

●4-65. How would you distinguish 16s from 23s rRNA, other than by size?

●4-66. List several lines of evidence that show that the genome of the RNA phages of *E. coli* is unique and not circularly permuted.

●4-67. In analyzing the sequence of a purified tRNA, an effective first step is to treat the tRNA with T_1 ribonuclease for a limited time at $0°$ C, in order to reduce the tRNA to two, three, or four large fragments which can be separated on DEAE (diethylaminoethyl cellulose) columns in 7 M urea.

(a) In this kind of experiment, how can one identify the 3'-terminal fragment?

(b) How can one identify the 5'-terminal fragment?

(c) The 3'-terminal fragment yields 7 Cp, 3 Ap, 1 methyl-Ap, 3 Up, 4 Gp, 1 A (no p), upon base hydrolysis. What information do these data give?

(d) Upon treatment with venom phosphodiesterase, the products are 1 methyl-A (no p), 4 pA, 7 pC, 3 pU, 4 pG. What information do these data give?

(e) Exhaustive hydrolysis with T_1 RNase yields 2 Gp and three oligonucleotide fragments, which yield, upon base hydrolysis: fragment 1, 2 Ap, 1A (no p), 4 Cp; fragment 2, 1 Gp, 1 Cp, 1 Up; fragment 3, 1 Gp, 2 Up, 2 Cp, and 1 methyl-Ap. Which fragment is 3'-terminal?

(f) Which fragment is 5'-terminal?

(g) Upon hydrolysis with alkaline phosphatase and venom diesterase, the following products are obtained from the T_1 fragments: fragment 1, 1A (no p), 4 pC, 2 pA; fragment 2, 1 C (no p), 1 pG, 1 pA, 1 pU; fragment 3, 1 methyl-A (no p), 1 pG, 2 pU, 2 pC. What are the 5'- and 3'-terminal nucleotides in each fragment?

(h) Treatment of fragment 2 with pancreatic RNase yields ApGp, Up, and Cp. What is the sequence of fragment 2?

(i) How would you complete the sequence studies on this 3'-terminal nucleotide, which is 19 bases long?

●4-68. (a) A pure tRNA species is degraded completely at 37° C with T_1 RNase, which attacks G residues and yields G-3'-P and oligonucleotides ending in . . . G-3'-P. The degradation products are separated by two-dimensional chromatography and electrophoresis. One oligonucleotide spot yields equimolar amounts of Ap, Cp, Gp, and Up after alkaline hydrolysis. What can be deduced from this information?

(b) This oligonucleotide is treated with alkaline phosphatase to remove 5'- or 3'-terminal phosphates and then degraded completely with venom diesterase to yield pA, pG, pU, and C. What can you deduce from this?

(c) When the original T_1 RNase fragment is treated with pancreatic RNase, Cp and Gp are released, along with a dinucleotide that releases Ap and Up after alkaline hydrolysis. What is the complete sequence of the original T_1 RNase fragment?

Note: Venom diesterase begins at 3'-OH ends and releases 5'-phosphoryl mononucleotides. Pancreatic RNase attacks pyrimidine residues, releasing C-3'-P, U-3'-P, and oligonucleotides ending in Cp or Up.

4-69. In *E. coli* infected with a ϕ80 phage carrying a *suIII* gene, you note on an electrophoretic gel a species of RNA slightly larger than $5s$. After carrying out an extensive T_1 nuclease digest, you observe a "T_1 fingerprint," which looks like one you have seen for a purified tyrosyl-tRNA (from a strain having the *suIII* marker). You investigate further by carrying out a limited T_1 digest at 0° C. One fragment A from this limited digest is treated further with T_1 and yields pppGp and a new fragment A'. Base hydrolysis of A' yields Gp, four Cp, and two Up. A' is also treated with phosphatase and then partially digested with snake venom diesterase. DEAE electrophoresis of this digest gives the pattern shown in Figure 4-7. Each spot is eluted from the DEAE thin-layer electrophoresis-plate and the base composition is determined after base hydrolysis, as indicated in Table 4-3.

FIGURE 4-7 ·

TABLE 4-3

Spot	Base composition		
1	2 Up	4 Cp	1 G
2	2 Up	3 Cp	1 C
3	2 Up	2 Cp	1 C
4	2 Up	1 Cp	1 C
5	1 Up	1 Cp	1 U
6	1 Cp	1 U	—

(a) What is the sequence of fragment A as far as you can tell?

(b) What does spot 7 of the limited venom digest represent?

(c) Would a pancreatic RNase digest of A′ have been useful in determining the sequence of fragment A? Why?

(d) RNA synthesis starts at the 5′-end of a molecule and proceeds to the 3′-end by the addition of nucleotide-5′-triphosphates. The sequence of tyrosyl-tRNA at the 5′-end is pGUGGU. . . . What might the significance of the pppGp nucleotide of the 5s RNA be? How might this 5s RNA be related to the tyrosyl-tRNA?

●4-70. Assume that a part of the single-stranded DNA of the filamentous phage M13 is being sequenced. The double-stranded replicative form is fragmented by a restriction enzyme. One fragment is isolated, denatured, and hybridized with the single-stranded DNA. The oligonucleotide bound to the single-stranded DNA serves as a primer for replication of a specific region of the DNA. This primer is extended by use of *E. coli* DNA polymerase I and low concentrations of deoxynucleoside triphosphates, for a limited time. The lengths of the products obtained by extension of the primer are in the range of 21–32 nucleotides.

The mixture of extended oligonucleotides is then divided into two portions. One portion is degraded by T4 DNA polymerase (using its 3′→5′ exonuclease activity) in the presence of a single deoxynucleoside triphosphate, which stops the exonucleolytic reaction short of the position corresponding to this added triphosphate. (For example, letters "+A, +G," and so on denote the products of T4 polymerase *degradation* in the presence of dATP, dGTP, and so on.)

The second portion of the mixture of extended oligonucleotides is further extended by *E. coli* DNA polymerase I in the presence of a mixture of only three deoxyribonucleoside triphosphates. (In this case, letters such as "−G" denote the products of the *extension* reaction in which dGTP is not present, and so on.) After extension or shortening, each mixture is electrophoresed on gels. The position of the bands is a measure of the size of each fragment, which is an indication of the number of nucleotides separating the primer from the nucleotide determining the size of the fragment (see Figure 4-8).

Read the DNA base sequence from the gels.

Size of fragment	+A	+G	+T	+C	−A	−G	−T	−C
32			—		—			
31	—						—	
30		—			—			
29				—		—		
28			—					—
27	—						—	
26		—			—			
25		—					—	
24				—			—	
23	—							—
22	—					—		
21	—					—		

FIGURE 4-8

4-71. The nucleotide sequence of a region of a single-stranded DNA molecule can be determined by the following procedure [see Maxam, A. M. and W. Gilbert, *Proc. Nat. Acad. Sci., U.S.* (1977), **74:** 560–564].

A sample of the DNA is separately given one of four different treatments.

1. The DNA is reacted with dimethyl sulfate, which methylates both G and A, though the reaction with G is five-fold faster. The DNA is then heated; the heating removes the methylated G and A residues. Then alkali is added and the sugar to which the methylated base was attached is cleaved from both neighboring phosphate groups.

2. After dimethyl sulfate treatment as in (1), the DNA is placed in dilute acid; this preferentially removes A. The DNA is then exposed to alkali to cleave the phosphates.

3. The DNA is reacted with hydrazine and then with piperidine. This results in sugar-phosphate breakage as in (1) and (2), but at sites of T and C, instead, *with equal frequency.*

4. Treatment (3) is carried out in the presence of 2 *M* NaCl; this suppresses the reaction with T.

These reactions are carried out such that only one base in fifty is altered by the methylation or hydrazine reactions. The hydrolytic steps are carried to completion. Each reaction mixture consists of a series of fragments whose sizes are determined by the location of particular bases. Electrophoresis of each mixture through a polyacrylamide gel separates the fragments by size. The DNA is usually terminally ^{32}P-labeled, so that the fragments in the gel can be located by autoradiography. The sequence is determined by observing the intensity of the bands found at each position. Note that the position tells where the base is but the intensity is a function of the probability of cleavage.

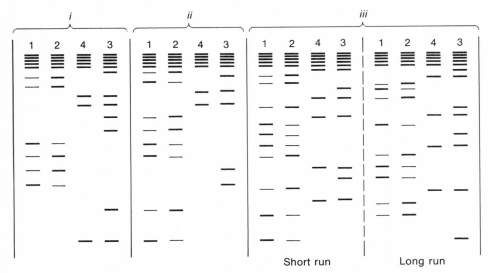

FIGURE 4-9

(a) Why is the initial reaction designed to terminate when one in fifty bases have reacted; in other words, why is one in two or one in one not chosen?

(b) What is the sequence of the two DNA samples shown in panels *i* and *ii*?

(c) A DNA molecule much longer than those in (b) can be sequenced. However, often all of the bands cannot be resolved on a single gel. To get around this, the reacted samples are divided into two parts, and each is treated by electrophoresis but for different lengths of time. What is the sequence of the DNA shown in panel *iii*?

4-72. You have isolated two lactose operon-transducing phage particles, one from phage λ (λ*lac*) and one from phage φ80 (φ80*lac*). Both phages contain unique DNA molecules, double-stranded, with cohesive ends; there is a very little homology between the

DNA molecules of the two phages. The physical maps of these phages are shown below.

The letters $i,p,o,z,$ and y indicate components of the *lac* operon. The numbers indicate arbitrary reference points in the phage DNA molecules. The primes (') indicate only that the *lac* operon and the references are in the DNA of the $\phi80lac$.

How could you use these phages to isolate pure DNA for the *lac* operon?

●4-73. Two cultures of the same *E. coli* K12 strain are grown in two different media: either broth, or a minimal medium containing only salts and glucose. Each culture is exposed to ^{32}P, at the same specific activity, for half a generation time. Messenger RNA is isolated from each culture by appropriate extraction from the polyribosome fraction (rRNA is discarded). Control fractions of unlabeled mRNA are also isolated. Each of the two ^{32}P-mRNA fractions is hybridized with *E. coli* DNA at 30 percent efficiency (which is considered good hybridization). Hybridization competition experiments are carried out. (See Table 4-4, in which + indicates that the indicated sub-

TABLE 4-4

DNA*	^{32}P-mRNA broth	^{32}P-mRNA minimal medium	mRNA* broth	mRNA minimal medium	Percent hybrid
+	+	−	−	−	30
+	−	+	−	−	30
+	+	−	+	−	0.3
+	+	−	−	+	10
+	−	+	+	−	10
+	−	+	−	+	0.3
+	+	+	+	+	0.3

* Added in great excess of ^{32}P-labeled material.

stance is present in the hybrid mixture, and − indicates absence of the substance.) The percentage of hybridization is shown in the right-hand column.

Interpret these data in terms of what you know about hybridization experiments and the growth of bacteria under different conditions.

5

DNA Replication

Introduction

In the early stages of the investigation of the mechanism of DNA replication, it was expected that all organisms would use the same replication mechanism. It rapidly became clear this was not so because the DNA in all organisms does not have a single form. For instance, some organisms contain single-stranded instead of double-stranded DNA. Furthermore the configuration of the DNA varies; there are single- and double-stranded *linear* molecules, single- and double-stranded *circles,* and double-stranded *supercoiled circles.* As a final complication, significant structural changes often occur during a cycle of replication; for example, a linear DNA molecule usually uses a circular intermediate, and a single-stranded circle commonly replicates by using both a double-stranded circle and a circle with a single-stranded linear branch as intermediates. Nonetheless, biologists have retained the hope that a universal enzymatic mechanism of *chain growth* may be discovered, though recognizing that since different enzymes are often capable of carrying out the same chemical reaction (for example, polymerases I and III of *E. coli*), a reaction catalyzed by a particular enzyme in one organism might employ a different enzyme in another organism. This proves to be the case—all DNA synthesis proceeds by addition of a deoxynucleotide-5′-triphosphate to the

3'-OH group at the terminus of a growing chain but, although most *E. coli* phages and plasmids use DNA polymerase III, there are several utilizing DNA polymerase I. It is not yet clear why the structure of DNA varies from one organism to another; however, the structural changes occurring within a particular replication cycle are usually necessitated by special features of the life cycle of the organism (for instance, the mode of packaging of viral DNA). In general, variations in the enzymes and protein factors needed to replicate a particular DNA molecule merely reflect differences in the mechanisms of chain initiation and of strand unwinding, although this is not always the case.

BASIC FEATURES OF DNA REPLICATION

DNA synthesis is catalyzed by several DNA polymerases, enzymes which form phosphodiester bonds between a free 3'-OH group at the growing end of a DNA strand and the 5'-P group of an incoming deoxynucleotide. All DNA polymerases copy a *template* strand which determines the base sequence of the strand being synthesized. No known DNA polymerase is able to lay down the first nucleotide onto the template strand—that is, they are unable to initiate chain growth; thus a *primer* is needed—an oligonucleotide fragment hydrogen-bonded to the template strand. In most systems examined to date, the primer is a short piece of RNA synthesized by an RNA polymerase; only in the case of rolling circle replication is there a DNA primer (a 3'-OH terminus of a previously synthesized DNA strand). Synthesis of a primer is called *initiation.* When *E. coli* chromosomal DNA is initiated, the RNA primer is synthesized by the same RNA polymerase used for ordinary transcription. However, certain plasmid and phage DNA molecules require other RNA polymerases, even though they replicate within *E. coli.* Initiation requires other factors as well—for example, the *E. coli dnaA* gene product and, perhaps, site-specific endonucleases—but the precise functions of the initiation factors are not understood at present.

Propagation of the replication fork requires more than a polymerase and a source of nucleotides. For instance, the proteins formerly called *unwinding proteins, DNA-binding proteins,* and *melting proteins* are all needed. In 1977 it became clear that proteins such as these may be sorted into three types. The true unwinding proteins are apparently responsible for unwinding the double helix and providing the polymerase complex with a single-stranded template. Two such unwinding proteins, *DNA unwinding enzyme I* and the *rep protein* have been isolated from *E. coli,* although the role of the former in DNA replication is not known. For both proteins the energy needed to unwind the helix comes from the hydrolysis of adenosine triphosphate (ATP). As the DNA is unwound, the single-stranded region is stabilized by the binding of the *helix-destabilizing (HD) protein.* This protein has the interesting property of binding tightly both to single-stranded DNA and to itself. Thus, as single-stranded DNA is freed from the double helix, HD protein molecules align sequentially along the strand, a phenomenon described as "walking down the helix." A third enzyme, *DNA gyrase,* apparently acts *ahead* of the replication fork and aids in the unwinding. DNA gyrase is not yet well understood but plays an essential role in the replication of some circular DNA molecules since it is capable of removing the superhelical twists that would be introduced into a circle by the unwinding of the helix.

The HD protein forms a continuous linear aggregate growing in the direction of movement of the replication fork. At the nongrowing end of this linear aggregate is DNA polymerase III, which apparently displaces monomers of the HD protein and moves along the single strand. Other factors seem to be involved in the actual propagation of the polymerase. Some of these, for example, the *dnaB protein, copolymerase III*,* and *replication factor Y,* hydrolyze nucleotide triphosphates to nucleotide diphosphates and inorganic phosphate; their role in replication is not known, though, nor are they used in all replicating systems.

DISCONTINUOUS REPLICATION

All known DNA polymerases synthesize DNA in a single direction—that is, by addition of nucleotide-5'-triphosphates to a 3'-OH end. This introduces a complication, because the two strands of double-stranded DNA are antiparallel—that is, the 3'-OH end of one single strand is located at the same end of the double helix as the 5'-P end of the other:

$$3'\text{-OH} \underline{\hspace{4cm}} 5'\text{-P}$$
$$3'\text{-OH} \underline{\hspace{4cm}} 5'\text{-P}$$
Parallel

$$3'\text{-OH} \underline{\hspace{4cm}} 5'\text{-P}$$
$$5'\text{-P} \underline{\hspace{4cm}} 3'\text{-OH}$$
Antiparallel

Hence if growth of both strands proceeded in the direction of movement of the replication fork, one of the growing strands would have a free 3'-OH group and the other would a free 5'-P group, which is not possible with the known polymerases. A solution to this paradox came

from the work of Reiji Okazaki, who found in the late 1960's that one of
the strands (the *leading strand*) grows in the direction of movement of
the fork but the other strand (the *lagging strand*) is replicated in short
sections that grow in the opposite direction from the movement of the
replication fork (see Figure 5-1). (These segments are widely known as

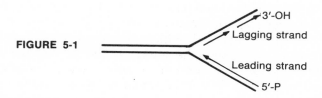

FIGURE 5-1

Okazaki fragments, though, as you will see, the term implies less difference between the leading and lagging strands than is now known to
exist.) The *discontinuous replication* allows synthesis to occur in two
directions within the replication fork yet maintains *overall* synthesis in
a single direction.

Since their discovery there has been considerable controversy about
the number, structure, and location of Okazaki fragments. It was first
thought that there are several fragments in a row in *both* newly synthesized strands, though in one strand these are not a necessary consequence of the directionality of the polymerase. However, it has
recently been shown that many of the fragments in the region of the
growing fork are not *precursors* of continuous strands of DNA but are
breakage products of longer strands resulting from excision of uracil
molecules that have accidently been incorporated into the growing
strands. The DNA polymerases do not efficiently distinguish
deoxyuridine triphosphate (dUTP) from thymidine triphosphate
(TTP). Thus since adenine can form a base pair with uracil as well as
with thymine, a DNA polymerase can insert uracil at a thymine site,
if dUTP is present. E. coli, and presumably other bacteria as
well, possess an enzyme, *dUTPase,* which converts dUTP to the
monophosphate dUMP (deoxyuridine monophosphate). Since a
monophosphate cannot be incorporated into DNA, this reduces the
possibility of accidental incorporation of uracil. However, some dUTP
molecules are not converted by dUTPase and are built into the growing daughter strand. These unwanted uracils are then removed shortly
after synthesis by an other enzyme, *uracil N-glucosidase,* which leaves
the deoxyribose in the DNA chain. Another enzyme, *apurinic endonuclease,* then removes the deoxyribose; by repair-polymerization,
thymine is inserted (presumably by polymerase I) into the appropriate
place in the chain. Apparently the insertion of thymine is not im-

mediate, so that fragments of DNA persist for a while. No term has yet been agreed upon for the fragments generated by the removal of uracil; in this book they will be called *uracil fragments;* we shall call a fragment generated by synthesis in the opposite direction from movement of the fork (that is, in the lagging strand) a *precursor fragment* (the term *nascent DNA* has been used in other scientific literature but since experimentally the word has meant *all* fragments in the replicating fork, precursor fragment seems to be a better term).

Because of the rapidity with which experiments are being carried out to examine the structure of the newly synthesized DNA in the replication fork, it is difficult at present to make a clear statement about events occurring in the fork. It is probably fair to say that although uracil fragments clearly exist, there is at present no unambiguous evidence for the existence of precursor fragments. The strongest evidence at present is that in an *E. coli* mutant that is unable to excise uracil some newly synthesized DNA is still in the form of short segments. However no evidence has yet been presented which proves that these segments are not derived from a larger piece of DNA. It seems necessary, though, that precursor fragments should exist in order to avoid the problem of the unidirectionality of the DNA polymerases. The most economical (but not necessarily the best) explanation of the observation of fragments in the replicating fork is the following: (1) The leading strand is synthesized continuously. (2) The lagging strand is synthesized as a precursor fragment. When a second precursor fragment is made, it is immediately joined to the first; a third fragment is then made and joined to the unit consisting of the first two joined fragments. Thus at any one time there is only a single incomplete fragment in the lagging strand. (3) Uracil is removed from *both* leading and lagging strands so that both strands are temporarily broken down into uracil fragments. (4) The uracil fragments are then joined by addition of thymine to form two continuous strands again.

Several early experiments by Okazaki also suggested that some of the fragments he observed are terminated by RNA. This was an attractive idea since a precursor fragment would probably have to be initiated by an RNA primer. However, these experiments have been shown to have been in error and reliable evidence for the existence of RNA is still much sought after. At the end of 1977, two tentative arguments could be put forth to account for terminal RNA on the precursor fragments: (1) the *dnaG*-gene protein, which is necessary for chain growth and probably also for synthesis of precursor fragments in *E. coli,* is an RNA polymerase; and (2) cells which lack the $5' \rightarrow 3'$ exonuclease activity of polymerase I accumulate a large number of fragments.

There are two aspects to the enzymology of DNA replication—initiation, and propagation of the replication forks. In some cases *in vitro* and *in vivo*, it is clear that RNA polymerase is involved in initiation. This enzyme alone is not sufficient for initiation, though, and other gene products (for example, the *E. coli dnaA*-gene protein), whose precise functions are unknown, are also required. The mechanism of propagation, which is to say, the events that take place at the replication fork, are also far from clear. A list of required gene products (all of whose functions are not known), including polymerase, ligase, unwinding proteins, and various DNA-binding proteins, has been made for the replication of the DNA of the phage ϕX174. However, different lists can be made for the replication of different plasmids, several phage DNA molecules, and the bacterial chromsome. For *E. coli* systems, some elements, such as polymerase III, the *dnaB*-gene protein, ligase, and binding proteins, are common to all lists.

New facts about the enzymology of DNA synthesis are rapidly being reported, and readers who are interested in keeping abreast of the most up-to-date information will find this each month in *Proceedings of the National Academy of Science*.

SEMICONSERVATIVE REPLICATION

Replication of double-stranded DNA is semiconservative, as was first demonstrated in the late 1950's by Matthew Meselson and Franklin Stahl. *Semiconservative replication* requires that the two parental strands separate in order to produce two daughter molecules. Each daughter molecule contains one parental strand. Before the experiments of Meselson and Stahl, unwinding was considered to be energetically so unfavorable that it was proposed instead that DNA replicates conservatively; *conservative replication* produces a single daughter molecule whose strands are both newly synthesized, and the strands of the parental molecule remain associated. (The parent molecule and the daughter molecule would, of course, then both serve as templates for the synthesis of two more molecules.) It was assumed that in conservative replication, local DNA bases could turn outward from the helix so that they could be copied by a polymerase; in this way the parental strand need not unwind and the winding of the strands of the daughter molecule would occur either base-pair by base-pair or a few base pairs at a time.

After the work of Meselson and Stahl it was suggested that DNA might be four-stranded, consisting of two double helixes associated by a new type of bond, the **biunial bond** (the nature of which was unspecified). Each double helix was assumed to replicate conservatively, thus avoiding the problem of the energetics of unwinding; the breakage of the biunial bonds between the parent duplexes and the reformation of these bonds between one parent and one daughter duplex makes the hypothesis consistent with the Meselson-Stahl work. This proposal was incorrect but the original work (described in Freifelder, 1978—see the Reference list) is interesting both for its logic and for the profound effect it had upon DNA research. It is no longer necessary to hypothesize a different DNA structure to explain the energetics of unwinding because it is now known that several unwinding proteins (for example, the *rep* protein) hydrolyze ATP and use the energy obtained from this hydrolysis to unwind the helix.

FOUR WAYS IN WHICH CIRCULAR DNA REPLICATES

Four distinct structures of replicating DNA have been seen by autoradiography and electron microscopy: the Cairns or theta (θ) structure, the replicating supercoil, the circle with D-loops, and the rolling circle or sigma (σ) structure.

1. The Cairns structure. In 1963, John Cairns obtained striking autoradiograms showing *E. coli* DNA replicating as a circle. Observation of a circular form made the unwinding of the strands seem even more unlikely than had been thought for linear DNA, and Cairns proposed that there should be a swivel or a swivel enzyme. The need for a swivel mechanism was made even more evident when Ross Inman and Maria Schnös showed that in some circular systems replication is bidirectional—that is, there are two replication forks moving in opposite directions around the circle. It is possible that a recently discovered enzyme, DNA gyrase, which is known to have a nicking–sealing activity, is the swivel enzyme sought. An alternative mechanism is that a break is introduced in one strand ahead of the replication fork, in order to allow free rotation, and is sealed at a later time by DNA ligase.

2. The replicating supercoil. Most supercoils, as, for example, in *E. coli* plasmid DNA, receive a single-strand break as part of the initiation step. This is often accomplished by a protein called the *relaxation protein,* which is bound to the supercoil and which, upon receiving some unknown signal, introduces a single strand break at a unique site in the DNA; this protein is then covalently linked to the freed 5'-P group. This generally leads to Cairns-type replication or sometimes to the rolling circle mode (described below). The DNA molecule of both polyoma and SV40 viruses is also supercoiled but is not converted to open circles before replication. Instead, the unreplicated template strands remain supercoiled while the progeny strands, which are not covalently linked to any parental strand, are contained in two enlarging untwisted loops. When replication is completed, these loops separate and are converted to supercoils. The intermediates are sometimes seen by electron microscopy as two loops on opposite sides of a tightly twisted region; hence the name *butterfly replication* is used occasionally. This is shown in Figure 5-2.

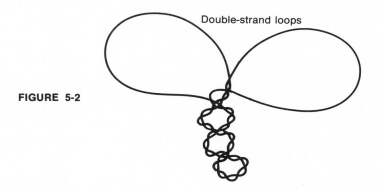

FIGURE 5-2

Double-strand loops

3. The circle with D-loops. In the early stages of replication of circular mitochondrial DNA, a loop is observed which consists of one single- and one double-stranded branch. This loop is a displacement-loop or *D-loop* (see Figure 5-3). The precise role played by the D-loop

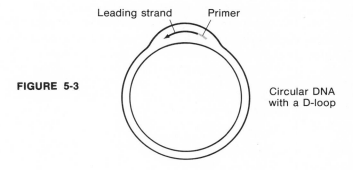

FIGURE 5-3

Leading strand   Primer

Circular DNA with a D-loop

in replication is not known but the following hypothesis has been given. It is assumed that replication begins with a priming event occurring on one strand. A single daughter strand (the leading strand) is extended from the primer, thus displacing the parent strand that is hydrogen-bonded to the strand being used as a template. Both parent strands are free of single-strand breaks and there is no swivel or swivel enzyme; therefore the unraveling of the helix at the replication fork causes twisting of the unreplicated portion. This twisting continues until no further twisting is possible, and hence, growth of the daughter strand stops. If growth stops before priming of the lagging strand has occurred (because the initiation sequence on the parent strand has not been reached), a loop is observed. Presumably, when the unreplicated portion has become fully twisted, a single-strand break is made, thus allowing free rotation that relieves the supercoiling, and then it is sealed. The leading strand is further extended until priming of the other parent strand occurs. Since replication of the second strand always lags behind that of the first, a single-stranded region is always observed. Note that three requirements must be satisfied before a D-loop may be seen. (1) There must not be a continuously acting swivel; (2) there must be a time delay before relaxation of the super-coiled region occurs; and (3) twisting must be maximal *before* priming of the lagging strand occurs (thus, the observation of D-loops is more probable in the replication of circles whose size is not much greater than that of a few precursor fragments).

4. The rolling circle. In late stages of phage-DNA replication and in DNA transfer from male to female *E. coli,* a linear DNA molecule is produced by replication from a circle. This is explained as follows. A single-strand break is made in one strand of the circle, generating a free 3'-OH group and a free 5'-P group. The free 3'-OH end is extended and displaces the parental strand bearing the 5'-P terminus. The displaced single strand is then converted to double-stranded DNA, presumably by means of synthesis and rejoining of precursor fragments. At this time the intermediate is a circle with a linear branch and is called a ***rolling circle*** or ***sigma molecule.*** (See Figure 5-4.) The

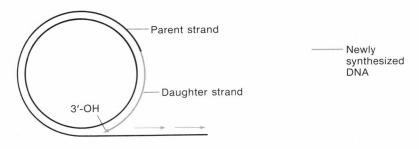

Rolling circle replication

three outstanding features of rolling circle replication are (1) the daughter strand is covalently linked to the broken parental strand, (2) the linear portion can grow indefinitely and become longer than the circumference of the parent circle, and (3) the unbroken parental circular strand is a master strand, serving as the source of the base sequence of both strands in the linear portion (except for the first unit replicated).

TERMINATION OF REPLICATION OF A CIRCLE

Termination of DNA synthesis refers both to completion of chain growth and to separation of the daughter molecules. Completion of chain growth could occur in two ways: by recognition of a particular base sequence in the parental strands or by collision of the replication fork either with the "start" terminus of the daughter strands (in the case of unidirectional replication) or with a second replication fork (in the case of bidirectional replication). Several *E. coli* plasmids have a termination sequence but for *E. coli* phage λ termination occurs wherever the two replication forks collide. With linear DNA molecules, separation of the daughter molecules occurs easily. However the replication of a circle yields two molecules which are linked as in a chain. At least one of the circles must be broken but in a way that prevents permanent conversion of the broken circle to a linear molecule. Presumably this is done by some enzyme possessing both nuclease and ligase activity.

The mechanism of termination is poorly understood.

REFERENCES

Alberts, B., and R. Sternglanz. 1977. Recent excitement in the DNA replication problem. *Nature*, **269:** 655–661. This is the most recent review of DNA replication. It is well illustrated, and readers will not require extensive background to understand it.

Freifelder, D. 1978. *The DNA Molecule: Structure and Function*. San Francisco. W. H. Freeman and Company. Contains an analysis of the four-stranded DNA model and its demise.

Kornberg, A. 1974. *DNA Synthesis*. San Francisco. W. H. Freeman and Company. The definitive book on the subject.

Stent, G., and R. Calendar. 1978. *Molecular Genetics*. San Francisco. W. H. Freeman and Company. Chapter 8 is an up-to-date introductory treatment with a historical approach.

Stryer, L. 1975. *Biochemistry*. San Francisco. W. H. Freeman and Company. Chapter 23.

Tye, B., P. Nyman, I. R. Lehman, S. Hochhauser, and B. Weiss. 1977. Transient accumulation of Okazaki fragments as a result of uracil incorporation into nascent DNA. *Proc. Nat. Acad. Sci., U.S.,* **74:** 154–157.

Watson, J. D. 1976. *Molecular Biology of the Gene.* Third Edition. Menlo Park, Calif., W. A. Benjamin. Chapter 9.

Problems

o**5-1.** State whether each of the following is true or false.

(a) In the synthesis of DNA the covalent bond which forms is between a 3'-OH and a 5'-P group.

(b) In general, the DNA replicating enzyme in *E. coli* is DNA polymerase I.

(c) A single strand of DNA can be copied if the four nucleotide triphosphates, and polymerase I, are provided.

(d) If polymerase I is added to the four nucleotide triphosphates without a DNA template, DNA is synthesized but with a random base sequence.

(e) An RNA primer must be complementary, in base sequence, to some region of the DNA.

o**5-2.** If one of the following enzymes is absent, not even one nucleotide can be added at the replication fork. This enzyme is

(a) Polymerase I (polymerizing activity).

(b) Polymerase I (5' → 3' exonuclease activity).

(c) Polymerase III.

(d) DNA ligase.

o**5-3.** What is the chemical group (for example, 3'-P, 5'-P, and so on) which is at the indicated terminus of the daughter strand of the extended branch of the rolling circle shown in Figure 5-5.

FIGURE 5-5

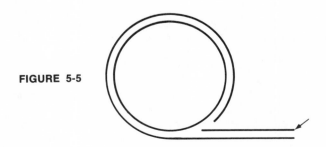

○5-4. What is the phenotype of each of the following *E. coli* DNA mutants: $dnaA^-$, $dnaB^-$, $dnaC^-$, $dnaE^-$, $dnaF^-$, $dnaG^-$, $polA^-$, $polB^-$, and lig^-, as far as DNA synthesis is concerned?

5-5. Consider the Meselson-Stahl experiment in which *E. coli* is grown for a long time in ^{15}N medium and then transferred to ^{14}N medium. (a) Assuming that the bacterial DNA is fragmented into at least 100 pieces during isolation and that replication is semiconservative, what fraction of the total DNA is found at the density of $^{15}N^{15}N$, $^{15}N^{14}N$, and $^{14}N^{14}N$ DNA at 0, $\frac{1}{2}$, 1, and $1\frac{1}{2}$ generations after transfer of the *E. coli* to ^{14}N medium? What would the distribution be if the DNA is not broken but remains as a single unit?
(b) Answer (a) for the case of conservative replication.

5-6. If, in a procedure such as that in problem 5-5, the data in Table 5-1 were obtained, what conclusion could you draw about the replication cycle?

TABLE 5-1

Generation	Percent of DNA		
	Heavy	Hybrid	Light
0	100	0	0
$\frac{1}{2}$	34	66	0
$\frac{3}{4}$	25	75	0
$\frac{7}{8}$	6	82	12
1	0	80	20

5-7. Consider 5-5 again. At 0 generations, when the medium is changed from ^{15}N to ^{14}N, a substance X is added. After approximately $\frac{1}{2}$ generation, 32 percent of the DNA is hybrid. After many hours, still only 32 percent of the DNA is hybrid. What does X probably do? (*Note:* The significance of the value of 32 percent may escape you. This is an actual value, though the main point that you should consider is that whereas *some* DNA becomes converted to hybrid the amount *never* exceeds this value.) Give an example of a type of substance that might behave like X.

5-8. Consider a Meselson-Stahl density-label experiment, as described in problem 5-5, in which cells are grown for a long time in ^{15}N medium and then transferred to ^{14}N medium for one generation. Assume that you are sure that the isolated DNA is two-stranded and that it is broken into approximately 100 fragments. If only $^{15}N^{15}N$ and $^{14}N^{14}N$ DNA are found (that is, no $^{15}N^{14}N$ is found), which of the following are true?

(a) DNA is not replicated by the Watson-Crick semiconservative scheme.

(b) DNA is semiconservatively replicated and the DNA is four-stranded.

(c) DNA is semiconservatively replicated and the DNA is two-stranded.

5-9. If a linear DNA molecule replicates with a Y-structure, from which properties of DNA and the *E. coli* DNA-polymerizing enzyme does it follow that both strands cannot be replicated by growth in the same direction (as shown in Figure 5-6)? How is this problem solved by the cell?

FIGURE 5-6

5-10. How is it shown experimentally that small fragments (Okazaki fragments) are present in the region of the replication fork?

5-11. What types of information have been provided to discriminate between the following two possibilities:

(a) Precursor fragments are synthesized on both sides of a replication fork.

(b) One strand is synthesized as a continuous unit and the other is synthesized discontinuously as precursor fragments.

5-12. To join together two precursor fragments, which of the following sequences of enzymes is probably used? Assume both fragments are already made.

(a) Pol I ($5' \rightarrow 3'$ exonuclease), pol I (polymerase), ligase.

(b) Pol III ($5' \rightarrow 3'$ exonuclease), pol III (polymerase), ligase.

(c) Ribonuclease, pol III, ligase.

(d) The *dnaG* protein, pol I, ligase.

5-13. What is the probable sequence of events used to join two uracil fragments?

5-14. At present even though there is no solid evidence proving that precursor fragments start with RNA, which of the following statements, if true, might be taken to support this idea? (More than one answer.)

(a) The *dnaG*-gene product is required for synthesis of a precursor fragment.

(b) Fragments are not joined if there is no pol I polymerizing activity.

(c) Fragments are not joined if there is no ligase activity.

(d) Fragments are not joined if there is no pol I ($5' \rightarrow 3'$ exonuclease) activity.

(e) After alkaline treatment some fragments can be labeled with ^{32}P by using the polynucleotide kinase reaction.

5-15. The gene *dnaE* codes for the polymerase that makes DNA in *E. coli*. The only known *dnaE$^-$* mutants are temperature-sensitive; that is, replication is normal at 30° C, but at 42° C, there is no DNA synthesis.

(a) Assume a culture of the cells is pulse-labeled with ^{3}H-thymidine at 30° C. Then the ^{3}H-thymidine is removed from the growth medium and at the same time the cells are shifted to 42° C. What happens to the precursor fragments during the period at 42° C?

(b) Would you expect to find radioactive precursor fragments in a *dnaE$^-$* mutant grown at 42° C in the presence of ^{3}H-thymidine?

5-16. RNA polymerase can only synthesize RNA in a single direction. Suppose you have a double-stranded RNA molecule that is being replicated by RNA polymerase. Would you expect to find precursor fragments of newly synthesized RNA?

5-17. Suppose you have a DNA molecule with a gap in one strand 5,000 nucleotides long and terminated with a 3'-OH group and a 5'-P group. If a DNA polymerase is added to this molecule *in vitro* (with everything else needed to make DNA), will the DNA filling the gap be a single piece or consist of short fragments? Would you expect fragments of any kind if the gap was filled *in vivo*?

5-18. In early studies of the properties of Okazaki fragments, it was shown that these fragments could hybridize to both parental strands of *E. coli* DNA. This was taken as evidence that they were synthesized on both branches of the replication fork. Although there is no rigorous evidence contradicting the idea that precursor fragments are found in both daughter strands, a particular characteristic of the replication of *E. coli* DNA invalidates a conclusion based solely on the hybridization data. What is this characteristic?

5-19. Consider a small, bidirectionally-replicating, circular DNA molecule known to utilize precursor fragments as intermediates. It is not known whether the fragments are in one or both branches of the replication fork. By electron microscopy, no single-stranded regions are seen; presumably they are too short. The replicating molecules are treated with an endonuclease active only against single-stranded DNA and are then observed by electron microscopy. Draw the structures which would be observed if the fragments were on either one side or both sides of the fork. Remember that, when isolated, all molecules will not be in the same stage of replication.

●5-20. The "Okazaki Piece" model for DNA replication is based upon the observation that a very brief labeling period with ^{3}H-thymidine produces incorporation of ^{3}H into DNA of low molecular weight. An alternative explanation is that this does not represent a precursor for ongoing chromosomal replication, but DNA of low molecular weight, which is rapidly turning over and is rapidly degraded if not used, and which is used as a *primer* for chromosomal replication (that is, it is used only at the beginning of a round of chromosome-replication but synthesized and degraded throughout the cell cycle if it is not incorporated into chromosomal DNA). Propose critical experiments that might make distinctions between these two models.

●5-21. Consider circular DNA the size of an *E. coli* DNA molecule. While replicating, a single swivel is sufficient to relieve the constraints imposed by the requirement for unwinding the helix during replication. Calculate the kinetic energy possessed by the rotating molecule if the swivel is made at a point 180° away from the origin of synthesis. Assuming that all of the energy produced in cleaving a triphosphate bond during nucleotide addition is used to rotate the DNA, is this amount of energy sufficient? Propose a way that this problem might be reduced.

5-22. Cytosine is frequently deaminated to yield uracil. However, the observed mutation frequency is not nearly as great as the deamination rate. Think about the cause of uracil fragments in the replication fork and try to complete an argument that leads to the conclusion that the high deamination rate has been the driving force behind the evolution of thymine, but not uracil, in DNA.

5-23. Which chemical reactions are catalyzed by *E. coli* polymerase I? By *E. coli* polymerase III? What are the different substrate requirements for the two enzymes?

●5-24. Suppose you are studying a DNA polymerase similar to *E. coli* polymerase I. This enzyme, like polymerase I, possesses a polymerizing activity and a $5' \rightarrow 3'$ exonuclease activity. It possesses other activities also but they are unknown to you. When the enzyme is in the presence of a double-stranded DNA template molecule containing a single-strand break with a free 3'-OH group and a free 5'-P group, and a mixture of the four radioactive deoxynucleotide 5'-triphosphates, radioactive DNA is detected. (This means that acid-insoluble radioactive material is found.) This is used as an indication of DNA synthesis since the nucleotides are soluble in acid and DNA is insoluble in acid. After 20 percent synthesis (that is, synthesis of an amount of acid-insoluble radioactivity equal to 20 percent of the weight of the template DNA), no increase in the molecular weight of the template DNA molecule is detectable and the density in CsCl remains that of double-stranded DNA. With a mutant enzyme, having normal polymerizing activity but lacking the exonuclease activity, after 20 percent synthesis both the molecular weight of the template DNA increases and the density of the radioactive material is greater than that of double-stranded DNA (but not as high as that of single-stranded DNA).

(a) What further property of this polymerase have you uncovered?

(b) What would happen to the density of the radioactive DNA if it is treated with an exonuclease that cleaves the molecule at the 5'-P terminus and hydrolyzes only single-stranded DNA?

(c) If the original DNA was radioactive and the deoxynucleotide triphosphates were nonradioactive, how would the amount of acid-insoluble radioactivity be affected by exposure to the wild-type enzyme and the four nonradioactive nucleotide triphosphates?

○5-25. Certain topological and energetic constraints result from the helical structure of DNA and the circularity of the template DNA molecule.

(a) Explain the roles of helix destabilizing proteins and unwinding enzymes in DNA replication.

(b) The *E. coli* ω protein can reduce the number of superhelical twists in naturally occurring supercoiled DNA. The DNA-gyrase enzyme can act on a nonsupercoiled DNA and introduce superhelical twists in the same sense as in naturally occurring DNA. However, only one of these proteins seems to be needed for DNA replication. Which is it and why?

5-26. A filamentous phage like M13 is isolated and called SCS13. The DNA of this phage is a single-stranded, circular molecule which cannot be replicated by the (pol III*-copol III*) complex in the presence of four deoxyribonucleotides. However, when the DNA-dependent RNA polymerase and four ribonucleoside triphosphates are added to the previously described mixture, the single-stranded DNA is replicated, and yields open, circular, double-stranded DNA containing a few ribonucleotides. In order to analyze the sequence of the initiating nucleotides you label all the deoxyribonucleoside triphosphates (dNTP) in the α-phosphate and let the ribonucleoside triphosphates remain unlabeled. The reaction product is treated with alkali and some ^{32}P is released, but only into guanosine-3'-P (not the deoxy form).

(a) What sequence information can be deduced from this result?

(b) If dATP is α-labeled with ^{32}P and no other dNTP is labeled, and if the label is transferred only to guanosine-3'-^{32}P, what new sequence information will be obtained?

(c) If you use ^{32}P-dATP as the only α-labeled triphosphate and no ^{32}P appears in free ribonucleotides, what sequence information can be deduced? Where does the ^{32}P label end up in this case?

●5-27. Phage K4 is a temperate phage with cohesive single-stranded ends on its DNA. In the phage particles, the DNA is linear, but it circularizes during its vegetative state, after infection. The replication-maturation cycle of K4 differs from that of λ in two ways: its DNA replicates as a simple Cairns circle and, in the normal phage assembly reaction, monomer circles are cleaved into the linear form subsequently found in phage particles. Suppose you are isolating mutants of K4 to study the replication mechanism and you identify three classes. Two of these (*dna* 1 and *dna* 2) are charac-

terized by recessive, nonsense, and temperature-sensitive mutations; nonsense mutations in these genes show no ^{3}H-thymidine
incorporation after infection of sup^- cells (cells not carrying a suppressor). The third class (*dna* 3) shows a low level of replication and
phage-particle production, has no nonsense or temperature-
sensitive mutations, and exhibits a partial *cis*-dominant defect in
mixed infection with wild-type phages (that is, more wild-type than
mutant phages are produced). Remarkably, if all recombination is
blocked by mutation, *dna* 3 mutants produce a smaller number of
phages in mixed infection with wild-type phages than by
themselves.

(a) Design a labeling and sedimentation experiment with
temperature-sensitive mutations to determine if the *dna* 1 and *dna* 2
genes affect the initiation or the propagation stages of replication.
Include a sketch of the sedimentation data.

(b) What do you think is defective in a *dna* 3 mutant?

(c) How can you explain the reduced growth in mixed infection
and the greater reduction when recombination is blocked?

●5-28. One problem in assigning a biological role to a known enzymatic
reaction is the lack of knowledge of the direction in which the
reaction goes *in vivo*. For example, DNA polymerase is usually
assumed to have a biosynthetic role in the addition of a nucleotide
to a 3'-OH terminus; however, the enzyme will also carry out
exonucleolytic degradation (at the 3'-end). Suppose *E. coli* phage λ
exonuclease could also function as a DNA polymerase, using at
least part of the same active site revealed by the nucleolytic reaction. What reaction would it be likely to catalyze? How would you
assay for such an activity?

5-29. Consider a phage particle containing a small linear, single-stranded
DNA molecule. Its replication mode is studied by centrifugation to
equilibrium in CsCl. Its density in CsCl is 1.714. Phages whose
DNA is ^{14}C-labeled infect the bacterium in nonradioactive medium.
Samples are taken at various times, DNA is isolated, and each sam-
ple is centrifuged. The results shown in Figure 5-7 are obtained.
How does this phage replicate its DNA? Do you think any progeny
phage will be ^{14}C-labeled?

5-30. Initiation of a round of DNA synthesis requires RNA synthesis. *E.
coli* RNA polymerase is inhibited by the antibiotic rifampicin. The
dnaG RNA polymerase is resistant to the drug.

(a) What is the consequence, with respect to DNA replication, of
adding rifampicin to a population of *E. coli* that is growing
logarithmically?

(b) What is the effect of rifampicin if the cells are first starved for a
required amino acid for 2 hours, and then both rifampicin and the
required amino acid are added?

5-31. *E. coli* DNA replicates bidirectionally. The required time is 40
minutes. Furthermore, it is known that at 37° C, even though more

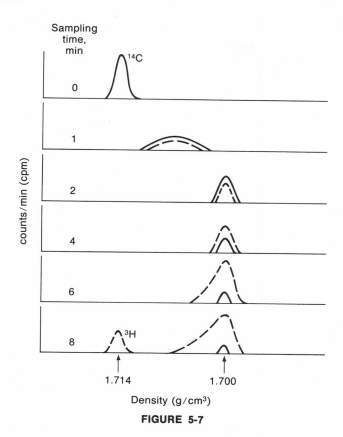

FIGURE 5-7

DNA is synthesized per unit time in a certain growth medium than in others, the rate of addition of nucleotides to the growing strand is the same in all growth media. Explain, then, how *E. coli* can divide every 22 minutes in some growth media.

5-32. Referring to problem 5-31, in some mammalian cells the rate of addition of nucleotides is about 5 percent of that in *E. coli*. How many growing points must there be in a mammalian cell containing 3 picograms of DNA per cell and replicating its DNA in 6 hours?

5-33. Which of the following are true? (Several are true.)
(a) There is no reason for a D-loop to be seen in a linear DNA molecule.
(b) If a circular DNA molecule receives a single-strand break before replication begins, it is unlikely that a D-loop will be seen.
(c) D-loops *must* enlarge by bidirectional movement of the replication forks.
(d) A D-loop can form without synthesis of a primer RNA.

●5-34. A mutagen NG has the property that when a bacterium is treated with it, double mutants which are very near one another are frequently produced. Distant double mutants are produced at very low frequency, as would be true of an ordinary mutagen. The data shown in Figure 5-8 can also be obtained with any strain of *E. coli*

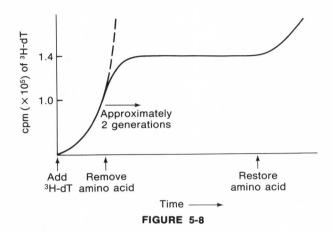

FIGURE 5-8

which requires an amino acid and which can take up ³H-thymidine (a specific label for DNA). The dashed line shows the data obtained if the required amino acid had not been removed.

A bacterium requiring nutrients A, B, C, . . . , Z is starved for a required amino acid for two generation times. Then the amino acid is restored and NG is added in two-minute pulses (ample time for mutagenesis). When tested for reversion, it is found that the reversion frequencies are increased about 100-fold; however, for any particular pulse only one gene shows the higher reversion frequency. Furthermore, the temporal sequence of the increases in reversion frequency follows map-order starting from an arbitrary point in the circular genetic map of *E. coli*. That is, if the map is circular, as shown in Figure 5-9, the time at which each gene *ABC*, . . . , shows the increase in reversion frequency can be plotted as in

FIGURE 5-9

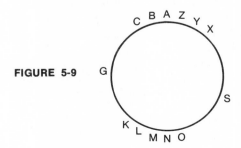

Figure 5-10. What do these results imply? That is, what special property does NG probably have? What does amino acid starvation do to the DNA replication cycle? What is the effect of readdition of amino acids? What do these experiments tell about DNA replication in *E. coli*?

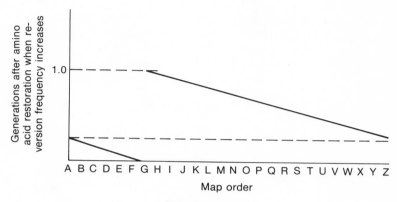

FIGURE 5-10

5-35. Will a donor DNA fragment, introduced into a recipient cell by transformation, be able to replicate in the recipient cell *without* integration? Explain.

5-36. A large circular DNA plasmid is digested with a restriction endonuclease. The fragments are allowed to reassemble at random (so that the fragments are no longer in the original order) and circles having the same molecular weight as the original plasmid are selected. These fragments are used to infect host cells lacking the plasmid. Assuming that ability to replicate is the only requirement for successful infection, what alterations in fragment order might prevent successful infection? If the order of the fragments is the same, but in some circles an additional fragment is added adjacent to an identical fragment, will this larger circle be capable of successful infection? Is it possible to delete a fragment, keeping the order otherwise the same, and still have infection? Explain.

5-37. Bidirectional replication was first detected during vegetative growth of *E. coli* phage λ by electron-microscopic observation of partially denatured DNA. A partially denatured linear DNA isolated from another phage particle has the denaturation map shown in Figure 5-11. Draw a partially denatured, replicating, circular DNA molecule that would be obtained after 50 percent of the molecule is replicated, if the molecule replicates unidirectionally

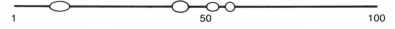

1   50   100

FIGURE 5-11

from a position exactly in the middle of the linear DNA. Repeat for bidirectional replication.

5-38. Consider the phage DNA in problem 5-37 and assume it replicates bidirectionally. In order to determine whether termination results from simple collision of the replication forks or by stopping of the forks at a genetically defined site, a phage DNA molecule is constructed in which the piece of DNA located 36–47 percent from the left end is duplicated in tandem sequence and the piece from 59–68 percent is deleted. The following molecules were observed. Is there a defined terminus?

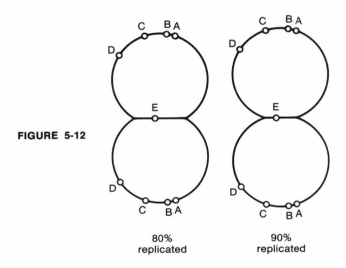

FIGURE 5-12

80% replicated   90% replicated

●5-39. What kinds of recognition sites on DNA are required for unidirectional replication, bidirectional replication, initiation using D-loops, and replication of a single-stranded circular DNA?

●5-40. Before bidirectional replication was known, one of the pleasing conceptual advantages of circular replicating molecules was that circularity allowed complete replication from a unique origin with a specific sequence; this provided for a specificity in initiation that starting at an end could not. Now that bidirectional replication has been established, we need to think some more about the significance of circular replicating molecules because linear molecules can replicate perfectly well in bidirectional fashion from a unique sequence. (T7 does!) Propose at least two "reasons" why a circular intracellular DNA molecule may be an advantage for a phage (even if the DNA is linear in the virus particle). Propose a way to test your suggestions.

5-41. Replication in phage ϕ19 DNA passes from an "early" phase in which circles generate new circles to a "late" phase in which very long molecules are produced. In the course of studying this phenomenon by using temperature-sensitive replication-mutants (phenotypes Dna1 and Dna2), and mutants defective in late transcription (reg⁻), you obtain the data shown in Table 5-2.

TABLE 5-2

Mutation	Replication (thymidine incorporation) after temperature increase	
	at 0 min	at 15 min
dna 1-ts	None	None
dna 2-ts	None	Normal
reg⁻	Normal	Normal
reg⁻ dna 2-ts	None	None

Propose a model for regulation of the early and late phases of $\phi 19$ DNA replication. What additional experiments would you conduct to try to test your proposal?

5-42. Read the following papers: (1) Cairns, J., *J. Mol. Biol.* (1963), **6:** 208–213; and (2) Cairns, J., *Cold Spring Harbor Symp. Quant. Biol.* (1963), **28:** 43–46.

In these articles, it was *assumed* that DNA replication in *E. coli* is unidirectional—the famous chromosome picture on page 44 of Cairns's second article is interpreted in this way.

(a) Explain Cairns's interpretation of how the labeling pattern was generated. (Diagram the chromosome starting at zero time of labeling.)

(b) Can you explain the labeling pattern shown in the insert (see page 44 of the second Cairns article) now that we know replication to be bidirectional?

(c) Take a hard look at the actual autoradiograph. Make your own diagram of doubly- and singly-labeled DNA stretches. Is your diagram consistent with bidirectionality?

(d) Design an autoradiographic experiment with synchronized bacterial cells to decide whether bidirectional synthesis terminates *(i)* at a unique place on the genome, or *(ii)* wherever the two forks happen to meet.

5-43. Read Helmstetter, C., S. Cooper, O. Perucci, and E. Revelas, *Cold Spring Harbor Symp. Quant. Biol.* (1968), **33:** 809–822.

(a) Explain how the data in Figure 1A of the article was obtained. In particular, how did the raw data for the culture with a generation time of 72 minutes look as a graph of *time of elution* versus *counts per minute (cpm) per cell eluted off of the filter?* (Helmstetter's earlier papers, cited in the foregoing reference, will describe in detail the synchrony technique used.)

(b) Make a chromosome diagram modeled after those shown in Figure 4 in the article, for cells growing with a 22-*minute* generation time at 37° C. (Assume C = 40 min, D = 20 min.)

●**5-44.** *E. coli* strain A has a temperature-sensitive *(dna-ts)* mutation in a gene involved in the initiation of replication; the strain is also *lac⁻*. At 42° C, no colonies can form when the bacteria are plated on agar. Strain B is wild-type and contains the transmissible sex factor F'Lac. Strains A and B are mated and cells having the genotype *lac⁺dna-ts* (strain C) are selected. These also fail to grow at 42° C, and segregate so that about 0.8 percent are *lac⁻ dna-ts* cells per generation. Although at 42° C, colonies from a culture of A arise at a frequency of 10^{-6}, temperature-resistant revertants of strain C arise at a frequency of 10^{-4}; these survivors have also lost the property of segregating *lac⁻* cells. Explain how C probably arose. Does C contain the *dna-ts* allele? Could this phenomenon occur for an *E. coli* strain that is mutant in either of the genes *dnaA* or *dnaE*?

●**5-45.** An *E. coli* strain is grown for five seconds in medium containing ³H-thymidine (³H-dT). The cells are chilled and broken open by ultrasonic vibration, a procedure which fragments the DNA to pieces approximately 5×10^5 daltons in size. If the broken cells are centrifuged, it is found that almost all of the radioactivity is at the bottom of the tube. If the extract is treated with ether or acetone, none of the radioactivity is at the bottom; it is all in the supernatant. If the cells are labeled for five seconds with ³H-dT and then grown for one minute in nonradioactive dT, none of the radioactivity is at the tube bottom. Explain.

●**5-46.** Replicating bacterial DNA is thought to be membrane-bound. M. Schaechter and his colleagues developed an interesting way to measure membrane binding. Bacteria are lysed by using procedures which neither destroy the membrane nor break the DNA. Crystals of magnesium sarcosinate are then added. These crystals bind the membrane fragments and, therefore, anything attached to a membrane. The mixture is centrifuged and the crystals sediment much more rapidly than unbound DNA. The fraction of DNA which is rapidly sedimenting is a measure of the fraction of DNA which is membrane-bound. It is of some interest to determine the number of binding sites on the DNA. This can be done by introducing double-strand breaks in the bacterial DNA (best done with X-rays) and obtaining a curve relating the amount of membrane-bound DNA as a function of the number of double-strand breaks. Draw the expected curves for 1, 2, and 3 attachment sites. (Remember to use the Poisson distribution.)

5-47. If *E. coli* is labeled for a few seconds with ¹⁴C-methionine, ¹⁴C is incorporated into the DNA. What does this represent, and what function does it serve?

6

Nucleases and Other Enzymes

Introduction

All cells contain **nucleases**—enzymes that catalyze the hydrolysis of nucleic acid. These are of a variety of types. Some are active against both DNA and RNA; others act against only DNA or only RNA. Nucleases may be further classified as active against both single- and double-stranded nucleic acid, or as active against only one of these. There are also nucleases which can digest an RNA only when it is hydrogen-bonded to a DNA strand. We may also distinguish between **exonucleases,** which can only remove a terminal nucleotide, and **endonucleases,** which cleave both terminal and internal phosphodiester bonds. There are nucleases which yield a 3'-OH- and a 5'-P-terminated nucleotide and those producing a 3'-P and 5'-OH terminus. Some nucleases can cleave only those bonds next to or between a particular pair of bases; others recognize extended sequences having particular symmetry properties (for example, the **restriction endonucleases**). There is also an interesting class of ATP-dependent nucleases. A few single proteins have several activities—an example is *E. coli* DNA polymerase I, which has two exonucleolytic activities.

Surprisingly, even though nucleases have been extensively studied, information about their biological function has been obtained for only a few; among these are (1) the nucleases used in repair synthesis, (2) the *E. coli* RecBC nuclease and the phage λ exonuclease, both of which play an unknown role in genetic recombination, (3) the restriction endonucleases, and (4) the RNase responsible for degradation of mRNA. (This is not an exhaustive list). Even more surprising is that many nucleases seem, under laboratory conditions, not to have a necessary function. For example, *E. coli* mutants lacking exonuclease I, exonuclease III, or endonuclease I seem to grow normally. Undoubtedly, in nature these nucleases have a function or they would have been lost by mutation long ago; however, these functions have yet to be found. Many nucleases probably remain to be discovered, since several nucleolytic functions are known which are not satisfied by a known enzyme—among these are the endonucleases which are required in all hypotheses about the mechanism of genetic recombination.

For the most part, in molecular biology and biochemistry, nucleases have been viewed either as interesting chemical entities whose chemical properties and mechanisms of action have yet to be elucidated or as tools that can be used in the laboratory for base-sequence determination, for changing the form of some nucleic acids (for example, for changing DNA supercoils to nicked circles), or for solubilizing unwanted nucleic acids in mixtures. Since usefulness as a tool has been the major interest in molecular biology, it is the point of view taken in most of the problems contained in this chapter.

Properties of selected nucleases are listed in Appendix B, page 238. That table will be useful while you work some of the problems in this chapter.

REFERENCES

There are no beginner's references on nucleases known to this author. More advanced, but not difficult references, are

Davidson, J. N. 1972. *The Biochemistry of the Nucleic Acids.* New York. Academic Press. Chapter 9.
Lehman, I. R. 1963. "The nucleases of *Escherichia coli*." *In* J. N. Davidson and W. E. Cohen (eds). *Progress in Nucleic Acid Research. Vol. 2.* New York, Academic Press. p. 84–118. A rather complete list of nucleases and their properties.

Problems

○**6-1.** What is the preferred substrate for the following nucleases? (a–e are *E. coli* enzymes.)
  (a) Exonuclease I.
  (b) Exonuclease III.
  (c) Two nuclease activities associated with polymerase I.
  (d) UV endonuclease.
  (e) Endonuclease I.
  (f) Micrococcal nuclease.
  (g) Pancreatic ribonuclease.
  (h) Ribonuclease H.
  (i) Apurinic acid nuclease.

6-2. Why is the activity of most nucleases inhibited by the addition of a chelating agent such as EDTA?

○**6-3.** Describe how phage *E. coli* T4 polynucleotide kinase can be used to label the terminus of a DNA molecule. What must be done to most naturally occurring linear DNA molecules in order to accomplish this?

○**6-4.** DNA polymerases have the job of forming phosphodiester bonds between 3'-OH and 5'-P groups. However, they cannot join together these two groups since the reaction requires activation. The polymerases therefore use nucleotide triphosphates containing a high-energy phosphate bond. However, DNA ligases are capable of joining together a 3'-OH and a 5'-P group. How do the ligases satisfy the requirement for activation energy?

○**6-5.** What do you think is the biological role of the *E. coli* restriction endonucleases?

●**6-6.** Some strains of bacteria excrete large amounts of nucleases. Propose several reasons why such excretion is useful to the bacteria. Why do you think that the mammalian pancreas secretes large amounts of ribonucleases and deoxyribonucleases? Human blood is rich in nucleases. What function might these nucleases serve?

6-7. Suppose you suspected that a linear DNA molecule is terminally redundant. If terminal nucleotides could be removed to produce complementary single-stranded termini, the molecule could circularize. Which nuclease or nucleases would you choose to perform this experiment?

6-8. Suppose you wanted to hydrolyze a DNA molecule totally to mononucleotides. Which enzyme or enzymes would you choose?

6-9. What do you think is the reason that most nucleases fail to work in 1 M NaCl?

●6-10. Pancreatic DNase is a potent nuclease, yet it rarely hydrolyzes DNA to mononucleotides. The fraction of resistant DNA decreases either with increasing enzyme concentration or if the initial DNA concentration is lower (even if the enzyme-to-DNA ratio is unchanged). If the enzymatic hydrolysis is carried out in dialysis tubing suspended in a huge buffer reservoir, the hydrolysis is complete. Explain.

●6-11. Suppose you are studying a DNA molecule which, by many standard criteria, is single-stranded—it reacts rapidly with formaldehyde, it shows no increase in optical density when heated and its sedimentation coefficient is markedly lower in 0.01 M NaCl than 1 M NaCl. However, when treated with an exonuclease known to be active against single-stranded but not double-stranded DNA and to act only at a 5′-P terminus, only one-third of the DNA is hydrolyzed to mononucleotides. What possibilities exist for the structure of this DNA molecule? An additional fact learned later is that if the DNA is boiled in 0.01 M NaCl and then cooled, it can be totally hydrolyzed by the exonuclease, whereas if it is boiled in 1 M NaCl, again only one-third is hydrolyzed. What can you say now about its structure?

6-12. The following two polyribonucleotides are treated with various enzymes as indicated. Predict the digestion products.
(a) ApCpGpCpUpUpCp: snake venom phosphodiesterase, spleen phosphodiesterase, ribonuclease T_1.
(b) ApUpC: snake venom phosphodiesterase, spleen phosphodiesterase, pancreatic ribonuclease.

6-13. You have isolated an RNA molecule that you wish to convert to a double-stranded form. No known phage RNA replicase is able to accomplish this. What sequence of enzymatic reactions could you use to prepare the double-stranded RNA?

●6-14. An enzyme, S1 nuclease, acts preferentially on single-stranded DNA.
(a) The weak activity against double-stranded DNA molecules increases with temperature even when the temperature is much less than that producing an increase in optical density. Explain.
(b) If a double-stranded DNA molecule has a single-strand break, S1 nuclease can break the phosphodiester bond opposite the single-strand break. Explain. How will this be affected by increasing temperature?
(c) S1 nuclease is active against a supercoiled DNA molecule. Why? Will it make a double-strand break or a single-strand break?
(d) If a double-stranded DNA molecule is denatured and renatured, it remains a relatively poor substrate for S1 nuclease. However if the DNA molecules isolated from a phage A are mixed

with DNA molecules from a phage B that contains a point mutation, and the mixture is denatured and renatured, approximately half of the renatured DNA molecules are broken by S1 nuclease. Why? Is a particular fraction of the renatured DNA subject to S1 activity? Which one?

(e) If DNA molecules isolated from a strain of λ phage maintained in a laboratory in the United States are renatured with DNA molecules of a λ phage which is presumably genetically identical to the first strain but obtained from a laboratory in France, approximately half of the renatured DNA molecules are broken by S1 nuclease. Why?

6-15. The enzyme terminal transferase is a DNA polymerase in the sense that it adds nucleotide-5'-P to a free 3'-OH group. What property does it have that the polymerases do not have? Explain how terminal transferase could be used to join together two DNA molecules.

6-16. Explain the role of DNA methylases in bacteria. Why does methylation normally occur in or near the replication fork? Why do you think that methylases have not evolved to act on free nucleotides? What is a possible role of RNA methylases?

6-17. An exonuclease is presented with single-stranded DNA. If solubilizes only a small fraction of the nucleotides. Give several possible explanations and propose a means to test each hypothesis.

7

Mutations and Mutagenesis

Introduction

The sequence of bases in the DNA of a gene (1) determines the amino acid sequence of the protein specified by that gene and (2) defines certain nontranscribed recognition sites. A base alteration which leads to a phenotypic change (a change of activity of the protein) is called a *mutation;* both an altered protein and an organism or cell containing a mutation are termed *mutant* and the process of creating a mutation is *mutagenesis.* A mutant protein invariably has either (1) a different conformation than the *wild-type* (that is, the original nonmutant) or (2) an alteration in a site on the surface of the protein involved in binding to other molecules (for example, the catalytic or active site of an enzyme).

KINDS OF MUTATIONS

Strictly speaking, a change in any DNA base is a mutation. However, if there is a base change without a phenotypic change, the mutation is often not recognized; such a base change is called a *silent mutation.* These are of two types: (1) those in which a codon change does not alter any amino acid in the protein and (2) those which result in an amino acid change (substitution, addition, or deletion) without affecting the activity of the protein. Silent mutations are sometimes detected accidentally by genetic recombination. For example, consider a protein having 100 amino acids in which, at positions 16 and 82, there is a leucine and an alanine respectively. In a genetic cross between two organisms both of which are phenotypically wild-type but each of which contains one of the silent mutations, an organism can arise which synthesizes a protein containing *both* silent mutations; if this protein with the dual amino acid substitution is inactive, the existence of two silent mutations will be recognized.

Some mutations affect binding sites on either the DNA or the RNA. A few well-known examples are mutations in operators, promoters, and ribosome binding sites. A mutation of this type is **cis-dominant**—that is, in a diploid, it affects only the particular DNA molecule containing it. For example, consider an operon consisting of a promoter P and two proteins A and B made from genes whose transcription is controlled by P. A diploid cell containing two DNA molecules having genotypes $p^+a^+b^-$ and $p^+a^-b^+$ will be phenotypically A^+B^+ since the presence of the promoter in each enables functional A and B proteins to be made. However, if the genotypes are $p^-a^+b^+$ and $p^+a^-b^+$, the p^- mutation prevents transcription of the a^+ gene, since p^- and a^+ are on the same DNA molecule, and the phenotype will be A^-B^+. It should be noted that a promoter mutation of this type has the effect of permanently turning *off* gene expression; other site mutations, such as in an operator, would, on the contrary, permanently turn the operon *on*.

That there are two types of mutations called promoter mutations is often not made clear. The example in the preceding paragraph is of a mutation *reducing* promoter activity. A promoter can also be *created* by mutation. For example, there may be a base sequence, perhaps within a gene, in which a single changed base can generate a new promoter. This would have the effect of permanently turning on the transcription of a set of genes whose expression is normally regulated by some repressor-operator system.

If a new promoter generated by a mutation is very strong (that is, if transcription is initiated with high frequency), the original gene in which the mutation occurred might lose most of its activity since transcription of that gene might seldom be completed. Activity could also be lost if the mutation alters the protein made by that gene. However, sometimes the activity of the gene is reduced only slightly by a promoter mutation.

The base changes which occur in DNA are of two types. Those which result in a substitution of one purine for another purine or one pyrimidine for another pyrimidine are *transitions*—that is, GC is replaced by AT, CG by TA, or vice versa. If a purine and a pyrimidine are interchanged, a *transversion* occurs—that is, GC is replaced by CG, GC by TA, AT by TA, and vice versa. These terms are useful primarily to classify mutagens.

Mutations which affect translation are of three types. (1) The mutation that results in the replacement of one amino acid by another is a *missense mutation.* This may or may not affect protein activity. (2) If a mutation creates a chain termination codon, it is a *nonsense mutation;* nonsense mutations almost always cause significant loss of activity. (3) Very frequently one or more bases are added or deleted. This is called a *frameshift mutation* because, unless the number of bases added or deleted is a multiple of three, the reading frame of the code is shifted, resulting in a totally new amino acid sequence. This almost always results in total loss of activity of the protein. The term *leaky mutant* is sometimes used. This refers to an altered protein whose activity has been reduced but not eliminated. Leakiness can result from either an amino acid substitution which still allows partial activity or from the residual activity of a fragment generated by premature chain termination. Frameshift mutants are rarely leaky. Leaky mutations can be used in genetic studies as genetic markers but are of little value in the elucidation of biochemical pathways, because the residual activity of the mutant is usually so great that steps in the pathway following the one requiring the mutant protein occur normally as if the mutation did not exist.

MUTAGENESIS

Base substitution mutations result from errors in replication or repair; a base is misread by a DNA polymerase so that a noncomplementary base appears in the newly synthesized strand. Mutations are usually made in the laboratory by exposure to *mutagens.* Some mutagens are *base analogues*—these are incorporated into DNA and have a relatively high error-frequency in base pairing. For instance, as a base analogue, 5-bromouracil substitutes for thymine; it normally pairs with adenine but occasionally pairs with guanine. Other mutagens react with bases in DNA and either permanently convert them to compounds which have pairing properties other than those of the original base or eliminate them entirely from the DNA. This class of mutagens includes radiation. Many, but not all, *intercalating* substances cause frameshift mutations. (An intercalating molecule binds to DNA by insertion *between* the planar rings of the bases—acridines are examples.) Possible mechanisms for frameshift mutation include base addition during replication or, more probably, induction of displaced base pairing of repeating sequences during genetic recombination. For example, in recombination within a region of DNA having the sequence GAAAACT, the sequences GAAACT (base deletion) or GAAAAACT (base addition) might result. Other mechanisms have also been proposed and there is at present little agreement about the mechanism by which frameshift mutations are produced.

REVERSION

Reversion refers to the reappearance of the wild-type phenotype in a population of mutants, and most mutants revert. *A revertant rarely has the wild-type genotype* (that is, the original base sequence), since secondary mutations can often restore activity to a mutant protein. A simple example can illustrate one way in which this might occur. Consider a hypothetical protein which is stabilized totally by one ionic bond between two amino acids (8 and 49) which are positively and negatively charged respectively. A mutation changes 8 to a negatively charged amino acid and thereby destroys the activity of the protein. A subsequent change in 49 to a positively charged amino acid could restore the activity by allowing the ionic bond to re-form. This is called reversion by a second site mutation or *intragenic suppression.*

In practice revertants cannot always be isolated. For instance, consider a bacterium having a mutation which makes it resistant to the antibiotic streptomycin—that is, it can form a colony on agar containing streptomycin. To find a streptomycin-sensitive revertant (one that will not grow on this agar) it is necessary to form colonies on agar lacking streptomycin and then to transfer each colony to agar containing the antibiotic, to determine which one does not grow. If revertants occurred at a frequency of 0.1 percent, this would be tedious but possible; however, since a more typical frequency is one revertant per 10^6 bacteria, this kind of testing is formidable. On the other hand, when a positive selection can be made, revertants can easily be found. For example, in a population of val^- bacteria (unable to grow on agar lacking valine) a val^+ revertant is easily isolated merely by putting from 10^6 to 10^7 bacteria on agar lacking valine and picking a colony that appears.

Reversion is often studied to determine the properties of mutants. For instance, if a mutant phenotype is the result of two mutations (a *double mutant*), the probability of reversion will be very low; it will be the product of the probabilities for reversion of a single mutation. A mutation might also be a deletion of many bases, in which case reversion does not occur. A nonreverting mutant will not necessarily be a deletion, since a frameshift mutation is very unlikely to revert. However, a frameshift mutant can be distinguished from a mutant produced by a large deletion (but not by a deletion of merely one or two bases) since it can be induced to revert by growth of the organism in the presence of a mutagen that produces frameshifts. That is, a frameshift mutation consisting of a single base addition can be reverted either by deletion of a nearby base or by addition of two (or five, eight, and so on) bases in the same region.

Single base substitutions can usually be identified since they can be induced to revert by exposure to base-analogue or chemical mutagens.

Reversion can also occur by the creation of *extragenic suppressors.* These are usually mutant tRNA molecules or amino acyl synthetases which can, at a certain frequency, introduce either an amino acid at the site of a chain-termination triplet or a substitution of one amino acid for another. This will be discussed in greater detail in Chapter 11.

REFERENCES

Davis, B. D., R. Dulbecco, H. N. Eisen, H. S. Ginsberg, and W. B. Wood. 1973. *Microbiology.* New York. Harper & Row. Chapter 11.

Drake, J. 1970. *The Molecular Basis of Mutation.* Holden-Day. This is probably the best and most complete book on the subject.

Stent, G., and R. Calendar. 1978. *Molecular Genetics.* San Francisco. W. H. Freeman and Company. Chapter 6.

Stryer, L. 1975. *Biochemistry.* San Francisco. W. H. Freeman and Company. Chapter 25.

Watson, J. D. 1976. *Molecular Biology of the Gene.* Third Edition. Menlo Park, Calif. W. A. Benjamin. Chapter 10.

Problems

o7-1. Consider a DNA molecule with the following structure:

A	T	G	C	A	T	T	A	A	T	G	C	C
T	A	C	G	T	A	A	T	T	A	C	G	G
		1		2	3		4		5	6		

(a) If during replication, the T at position 2 is read as an A, what are the results of the first and second rounds of replication?

(b) If the replication apparatus copies the AT pair at 2 and then copies the AT at 5, skipping 3 and 4, what are the results of the first and second rounds of replication?

7-2. Spontaneous mutations (those which arise without requiring exposure to a mutagen) are very frequently frameshift mutations. Assuming that the base-recognition systems are not always perfect, explain how a frameshift mutation could arise by error in either replication or recombination.

7-3. Propose a mechanism for the production of a deletion mutant in which (i) one base pair is deleted and (ii) 1,000 base pairs are deleted.

7-4. *E. coli* polymerase I possesses several enzymatic activities. Two important activities are the polymerizing function and the $3' \rightarrow 5'$ exonuclease. Mutant polymerases have been found which either increase or decrease mutation rates in an organism containing the mutant enzyme. The mutant that increases the mutation rate is a *mutator;* the mutant that decreases the mutation rate is an *antimutator.* The mutator and antimutator activities are usually a result of changes in the ratio of the two enzymatic activities described above. How do you think the ratios change in a mutator and in an antimutator?

o7-5. Explain the basis of mutation by the following mutagens: 5-bromouracil, 2-aminopurine, nitrous acid, and acridine orange. Also state whether they induce transitions (for example, an AT pair replaced by a GC pair or vice versa), transversions (for example, an AT pair replaced by a TA pair), or frameshifts.

o7-6. Write down the codons which, if altered by a single base change, could give rise to chain termination.

o7-7. Under what circumstances might the appearance of a chain termination codon *not* result in loss of activity of a protein?

o7-8. Explain why all base-pair changes in DNA do not give rise to a protein with an amino acid substitution.

o7-9. Why does the substitution of a single amino acid not always produce a protein with reduced activity (that is, a mutant protein)? What kinds of amino acid changes would most likely result in a mutant protein?

o7-10. Which of the following might be involved in the mechanism of reversion of a single point mutation?

(a) The original amino acid is restored.

(b) The wrong amino acid is replaced by a second amino acid, different from the first.

(c) An amino acid elsewhere in the protein is changed.

7-11. Consider a bacterial gene containing 1,000 base pairs. As a result of treatment of the bacterium with a mutagen, mutations in this gene are recovered at a frequency of one mutant per 10^5 cells. One of these mutants is taken and a pure culture of this mutant is grown up. This culture is then treated with the same mutagen and revertants are found at a frequency of one per 10^5 cells. Would you expect the gene product obtained from the revertant to have the same amino acid sequence as the wild type? Explain.

7-12. Suppose a bacterium is treated with a chemical that alters a particular adenine at a single site in a gene. Because of this, the adenine appears to the DNA and RNA polymerases to be a guanine. Clearly after several rounds of replication of the DNA there will be a DNA molecule with a GC pair at the site where there was an AT pair. However, let us consider the bacterium before any replication occurs. Two requirements must be met if the bacterium has lost the ability to make a functional gene product from that gene. What are they?

○7-13. Nitrous acid, HNO_2, induces mutations by deaminating cytosine and adenine, thereby converting them respectively to uracil and hypoxanthine (which pairs like guanine). Which of the following amino acid changes could be induced by nitrous acid? Arg → His; Leu → Ser; Tyr → His; Met → Ile; Gly → Asp.

●7-14. Suppose you isolated a mutant bacteriophage that was capable of forming a plaque on bacterial strain A but not on strain B. How would you go about deciding whether you had a single point mutant, a double mutant, or a deletion? First answer the question assuming that it is the only mutant that has ever been isolated in this phage; then answer it assuming that you have a very large collection of mutants which have already been mapped.

7-15. One hundred *lac*⁻ mutants are examined separately to measure reversion frequencies. Of these, 45 revert at a frequency of 10^{-5}; 49 at 3×10^{-6}; 3 at 3×10^{-11}; and 3 at 10^{-10}. The three which revert at 10^{-10} are
(a) Single point mutants.
(b) Amber mutants.
(c) Probably double mutants but possibly deletions.
(d) Probably double mutants and definitely not deletions.

7-16. Suppose you are testing the idea that for each segment of DNA, both strands of DNA are used as a template for synthesis of mRNA. Which of the following results could be taken as evidence to support this idea? After answering the question, explain your answer.
(a) Mutations occur in pairs in that frequently two amino acids in a mutant protein are changed.
(b) Mutations are paired in that frequently two different proteins are simultaneously affected.

7-17. Assuming the hypothesis in 7-16 is correct and assuming that chain termination occurs in the standard way using only the known chain termination mutations, could two proteins be prematurely terminated by the same change in base pair? Explain.

(a) Always.

(b) Never.

(c) Sometimes.

●7-18. How would you show genetically that the reversion of a frameshift mutation occurred within a gene rather than by a second site mutation that occurred far away on the genome (for example, by a tRNA frameshift suppressor)?

7-19. Usually mutations are **recessive**. That is, in a diploid organism if a mutant and wild-type allele are both present, the organism is phenotypically wild-type. Give five examples of **dominant mutations**.

7-20. A bacterial mutant in the z gene of β-galactosidase is isolated. How would you identify genetically whether it is a frameshift, missense or nonsense mutant? If it is a nonsense mutant, how would you tell if it is an amber, an ochre, or an opal mutant?

7-21. Often dyes are incorporated into agar in order to determine whether a bacterium can utilize a particular sugar as a carbon source. For instance, on EMB agar containing a sugar X, suppose that a bacterium which has genotype x^+ makes a purple colony and one which has genotype x^- makes a pink colony. This would be because fermentation of the sugar has produced acid which has changed the dye to a purple color. If you had EMB agar containing lactose, xylose, and maltose, what would be the color of the colonies produced by bacteria having the following genotype? $lac^+xyl^+mal^+$, $lac^-xyl^-mal^+$, $lac^+xyl^-mal^+$, $lac^-xyl^+mal^-$, $lac^-xyl^-mal^-$?

7-22. Refer to question 7-21: If a population of lac^+ cells is treated with a mutagen that produces lac^- mutants and the population is allowed to grow for many generations, one finds on EMB agar plates a few pink colonies among a large number of purple ones. If the mutagenized population is plated on the agar immediately after it has been treated with the mutagen, some colonies appear which are called *sectored*—they are purple on one side and pink on the other side. Explain.

●7-23. How might you show whether a mutator gene can induce mutations in replicating and nonreplicating DNA? The following article will be helpful: *J. Mol. Biol.* (1970), **50:** 129.

7-24. How does one usually select for auxotrophic mutants in bacteria? Devise a scheme to isolate temperature-sensitive amber suppressors.

●7-25. Name two methods for obtaining mutants in a particular region of the *E. coli* genome?

●7-26. How does the use of hydroxylamine to generate ochre mutants of *E. coli* phages that have single-stranded DNA identify the strand of the double-stranded replicative intermediate that is transcribed? The following article will be helpful: *Nature* (1970), **228**: 54.

●7-27. Suppose you inoculate 200 tubes with about 50 bacterial cells each and incubate the cultures until each contains a few million cells. For 100 of the cultures, each culture is then placed on a separate plate loaded with phage T1. The other 100 cultures are plated on plates loaded with T4. After overnight incubation, 95 of the T1 plates and 5 of the T4 plates have some colonies.

(a) By what factor is the rate of mutation to T1 resistance greater than that to T4 resistance?

(b) Suppose 100 of the cultures were plated on plates loaded with both T1 and T4 phages. How many of those plates would have colonies after overnight incubation?

●7-28. Several hundred independent missense mutants, altered in the A protein of tryptophan synthetase, have been collected. It was hoped that at least one mutant for each of the 286 amino acid residue positions in the protein would be found. However, fewer than 30 of the residue positions were represented with one or more mutants. Suggest in a few sentences some possibilities to explain why this set of missense mutants was so limited.

●7-29. Biological membranes are composed mainly of lipid and protein. About one-third of the dry mass of the *E. coli* cell membrane is lipid, and all of this is phospholipid. One approach to delineating the roles of membrane phospholipids is through the isolation of mutants defective in the biosynthesis of various species of membrane phospholipids.

Describe a procedure for selection of *E. coli* mutants which are temperature-sensitive for synthesis of cell-membrane phospholipids. Use L-glycerol-3-phosphate (G-3-P) as a phospholipid precursor. As a starting strain, you are given *E. coli* that is constitutive for uptake of G-3-P and that cannot ferment G-3-P, owing to mutations which destroy the activities of alkaline phosphatase and G-3-P dehydrogenase.

●7-30. Would you expect the mutator activity of polA (polymerase I) to act in the *trans-* configuration? How can this be demonstrated? The following article will be helpful: *Mol. Gen. Genetics* (1975), **141**: 251.

●7-31. The spontaneous forward mutation rate is approximately 10^{-9} base pairs per generation. Suppose this mutation rate represents the error frequency in DNA replication. Calculate the approximate mutation rate if the error frequency depended solely on hydrogen-bonding energetics, assuming a GC base pair provides 3 kcal/mole and an erroneous GT pair provides 2 kcal/mole. Give at least two reasons why the mutation frequency might be so much less than the number you have just calculated.

●7-32. Assume that you can separate and fractionate the *r* from the *l* strand of T7 DNA and that each strand is infectious in *E. coli* spheroplasts (that is, each gives rise to phage particles by transfection). If each strand is separately mutagenized before transfection, discuss the likelihood of the generation of the following mutants by hydroxylamine.

(a) Ochre mutants from mutagenized wild-type *r* and *l* DNA strands.

(b) Ochre mutants from mutagenized *r* and *l* DNA strands that already carry opal or amber mutations.

Indicate the original and altered codons in the DNA (in the single-stranded and double-stranded state) and in the RNA.

●7-33. Phage Q3 is known to transduce *leu*. A *leu*⁺-transducing lysate is prepared and then mutagenized with nitrous acid. It is then used to transduce a *leu*⁻ strain at 30° C. Many of the *leu*⁺-transductant colonies are small and fail to grow when replated at 42° C. When a similar transduction is performed with an unmutagenized phage preparation, the *leu*⁺ colonies are more uniform in size and all grow at 42° C. Explain this phenomenon.

8

RNA Synthesis, Transcription, and Regulation

Introduction

Information flows from DNA to protein synthesis through a stage in which messenger RNA (mRNA), an RNA molecule which is complementary to one of the DNA strands in a gene, is synthesized. It is an obvious economy if cells synthesize proteins only when needed; hence various regulatory mechanisms have evolved, of which the most common is the switching on and off of mRNA synthesis.

PROCESS OF RNA SYNTHESIS

The enzymes responsible for the synthesis of RNA are the RNA polymerases. The one best understood is the *E. coli* RNA polymerase, whose properties are summarized here. This enzyme joins a ribonucleotide triphosphate to a 3′-OH group at the terminus of a growing polyribonucleotide to form an alternating copolymer of ribose and monophosphate. The base sequence of the RNA is complementary to that of the DNA template strand, but with uracil, not thymine, pairing with an adenine from the DNA. For any given segment of double-stranded DNA, usually only a single strand is copied. *E. coli* RNA polymerase is a multimeric enzyme consisting of five subunits called

β, β', α, ω, and σ. The $\beta\beta'\alpha\omega$-aggregate is called the *core enzyme* and is responsible for the actual polymerization. Without the σ subunit, the core enzyme is incapable of *initiating* RNA chains. *In vitro,* the core enzyme can occasionally initiate chains without the σ subunit, but this initiation often occurs at base sequences other than the correct start sequences recognized by the σ subunit. When the σ subunit is present, the enzyme is called the *holoenzyme* and it is then capable of binding very tightly to specific initiation regions of the DNA called *promoters.* A promoter is a sequence of bases which can be divided into several subsequences. Under the direction of the σ subunit, the holoenzyme binds first to the *RNA-polymerase recognition site* and then diffuses to the *RNA-polymerase binding site.* During this interaction, a localized unwinding of the DNA helix occurs and, in a way that is poorly understood, the first RNA base is hydrogen-bonded to a DNA base about 6 or 7 bases away from the binding site. As soon as the first phosphodiester bond is formed, or shortly thereafter, the σ subunit dissociates from the core enzyme, but chain elongation continues until a termination signal is reached. After termination, the core enzyme, the DNA, and the RNA detach from one another and a σ subunit binds to the core enzyme to enable the cycle to get started again.

There are several types of start signals: promoters do not all have the same base sequences nor do they have the same affinity for RNA polymerase. Presumably the differences are a regulatory mechanism for controlling the relative amounts of different mRNA molecules which are being synthesized. Stop signals are of two types—those which are recognized by RNA polymerases alone and those which work only in the presence of a protein called *rho* factor. Preliminary data suggests that *rho* factor is an enzyme that cleaves ribonucleotide triphosphates to monophosphates, but it is likely that the mechanism by which synthesis terminates is not simply one by which the availability of triphosphates is reduced. The mechanisms of initiation and termination of transcription are subjects of intense study at the present time.

REGULATION OF TRANSCRIPTION

Regulation is necessary to avoid unnecessary synthesis. The simplest type of regulation is that found in the *negatively controlled* operons such as the *E. coli lac* operon. In the general case, the operon consists of four elements: a *repressor* gene, *operator* and *promoter* sites, and the *structural genes* coding for the proteins whose synthesis is regulated. The repressor protein is synthesized at a relatively constant rate and is capable of binding tightly to the operator sequence. When it is on the operator, the repressor is positioned in such a way that it shields the promoter from RNA polymerase, so that transcription cannot be initiated. Thus, as long as an active repressor is present, the operon is not transcribed. When the enzymes of the operon are needed, a cell must have some way to initiate their transcription. This is most easily done by binding the enzyme substrate (or a related product) to the repressor in such a way that the repressor no longer recognizes the operator; one can easily imagine (see Chapter 2) how binding a small molecule to a repressor can change the shape of the repressor molecule and thereby eliminate the site where it binds to the operator. A molecule that can eliminate the operator-binding activity of a repressor in this way and thereby dissociate a repressor-operator complex is called an **inducer.** Once the repressor is removed from the operator, RNA polymerase can bind to the promoter and mRNA can be synthesized. The enzymes of the operon will then be made and these will act upon the substrate. Their action will continue until the substrate is consumed. If the substrate or a closely related substance is the inducer, repression will be reestablished when the inducer is no longer present. That is, as the inducer concentration decreases, the equilibrium

$$\text{Repressor (R)} + \text{Inducer (I)} \rightleftarrows \text{RI Complex}$$

will shift to the left and the active repressor will increase in concentration until it can bind to the operator and thus prevent further binding of RNA polymerase. In this elegant way, the proteins of the operon are made only when needed.

Some negatively regulated operons use the *end-product* of a biosynthetic pathway as a **corepressor** rather than a substrate as an inducer. An example is the operon in *E. coli* responsible for synthesis of the amino acid tryptophan. If tryptophan is present in the growth medium of the bacterium, there is no need to have this operon turned on; if tryptophan is absent, the operon must be switched on if the cell is to be able to synthesize proteins. This regulation is accomplished by means of a repressor that is inactive unless a corepressor (in this case, tryptophan) is bound to it. Thus the repressor is active and bound to the operator only when tryptophan is present. If tryptophan is depleted from the growth medium, the tryptophan-repressor complex will dissociate, the repressor will be removed from the operator, and transcription of the operon will begin. Note that regulation using the end-product as a corepressor is efficient when the role of the operon is to *synthesize* an essential molecule and that the use of a substrate as an inducer is of value when the operon specifies a *degradative* pathway.

Operons that consist of several proteins usually make a polycistronic mRNA, which is a single RNA molecule that contains the codon sequences for each of the proteins of the operon. The proteins are then translated from this single mRNA molecule through implementation of appropriate start and stop signals. Surprisingly, the number of protein molecules of each type translated from a single polycistronic mRNA is not the same—some proteins are made more frequently than others. The mechanism by which this translation is regulated is not well understood, but it probably has something to do with an effect of the base sequence that is adjacent to the AUG "start"-codon on the efficiency of initiation of translation.

Many operons are regulated by positive control elements instead of by repressors. Positive control elements can act either by binding directly to the promoters, to proteins which bind near promoters, or to RNA polymerase itself. Most positively-regulated operons are relatively complex, but the principle of all is basically the same: the substrate (or a closely related compound) of an enzyme of the operon stimulates the interaction of RNA polymerase and the promoter by a steric alteration of some kind.

As more operons are being studied it has become clear that many operons are under dual control. A simple example is a class of operons responsible for sugar metabolism. These are regulated both by the sugar to be metabolized and by glucose. If both glucose and another sugar are present, there is no need for a cell to synthesize the enzymes responsible for utilizing that sugar, since the enzymes for glucose utilization are always present (the final stage in the catabolism of almost all carbohydrates is glucose degradation). Therefore it might be expected that glucose would have the ability to regulate transcription of a sugar operon even if the sugar operon is derepressed. Derepression is achieved through the *catabolite activator protein* (CAP), which binds to specific sites in DNA near the promoter for the sugar-utilizing operons only when cyclic AMP is bound to CAP. Thus for the CAP-dependent operons, initiation of transcription requires both derepression and a CAP molecule that has been bound to the DNA. The synthesis of cyclic AMP is regulated by glucose metabolism. When glucose is present, the concentration of cyclic AMP is low, CAP cannot bind to the binding site on the DNA, and transcription of the CAP-dependent sugar operons cannot occur. Thus, if both glucose and, for example, lactose are present, glucose is selectively utilized. Once the glucose is consumed, the concentration of cyclic AMP increases, CAP binds to DNA, transcription of the *lac* operon begins, and the lactose is utilized. Furthermore, if glucose is added to a cell which is catabolizing lactose, the lactose is no longer needed, transcription of the operon will rapidly be turned off, and the cell will shift to glucose utilization.

REGULATORY MUTANTS

The study of operon function has been made possible principally by
the isolation of mutants in the regulatory elements. Since regulation
basically involves the interactions between a repressor, operator, and
an inducer or corepressor, a variety of mutations of different types are
possible.

In an inducible operon, the repressor has two binding sites (one for
the operator and another for the inducer), and thus there are four kinds
of single mutations that may occur: (1) the inducer binding site is
inactivated so that derepression can never occur; (2) the inducer binding site is strengthened so that once induction occurs, repression cannot be reestablished; (3) the operator binding site is inactivated so that
the repressor can never bind to the operator; and (4) the operator
binding site becomes strengthened so that the repressor cannot be
removed from the operator even when the inducer is present.

In operons having corepressors there may also be four kinds of repressor mutations: (1) the corepressor binding site is altered so that the
repressor cannot bind the corepressor and thus repression never occurs; (2) the corepressor binding site is strengthened so that repression
is not eliminated when the concentration of the corepressor decreases;
(3) the operator binding site is modified so that the repressor can bind
to the operator even in the absence of the corepressor; and (4) the
operator binding site is modified so that the repressor can never bind
to the operator.

The operator may be mutated also, in three ways: (1) to prevent
repressor binding; (2) to have such tight binding that the repressor
cannot be removed even when the inducer is bound; and (3) the mutation causes the operator to induce a conformational change in the repressor so that the bound repressor is no longer recognized by the
inducer. By construction of partial diploids harboring various regulatory mutants, the different mutant types can be identified and dominance relations can be established. This is explored in many of the
problems in this chapter.

REFERENCES

Davis, B. D., R. Dulbecco, H. N. Eisen, H. J. Ginsberg, and W. R. Wood, 1973.
  Microbiology. New York. Harper & Row. Chapter 11.
Jacob, F. and J. Monod, 1961. "Genetic regulatory mechanisms in the synthesis of proteins." *J. Mol. Biol.*, **3**: 318–356. In this paper the operon model
  was first described.
Lehninger, A. L. 1975. *Biochemistry*. New York. Worth Publishers, Inc. Second edition. Chapters 33 and 35.

Stent, G., and R. Calendar. 1978. *Molecular Genetics*. San Francisco. W. H. Freeman and Company. Chapters 15 and 20.

Stryer, L. 1975. *Biochemistry*. San Francisco. W. H. Freeman and Company. Chapters 24 and 27.

Watson, J. D. 1976. *Molecular Biology of the Gene*. Third Edition. Menlo Park, Calif. W. A. Benjamin. Chapter 11.

Problems

o8-1. Write down the two RNA sequences which could conceivably result from complete transcription of the following DNA duplex.

5′	A	G	C	T	G	C	A	A	T	G	3′
3′	T	C	G	A	C	G	T	T	A	C	5′

Indicate the 5′ and 3′ ends of each transcript.

o8-2. Which of the following are steps in RNA synthesis?
(a) Binding of DNA polymerase to DNA.
(b) Binding of RNA polymerase to DNA.
(c) Binding of a σ factor to RNA polymerase.
(d) Binding of a σ factor to DNA polymerase.

o8-3. Which of the following are synthesized directly from a DNA template?
(a) Messenger RNA.
(b) Transfer RNA.
(c) Repressor.
(d) Ribosomal RNA.
(e) Protein.

o8-4. Define the following terms: RNA polymerase core enzyme; RNA polymerase holoenzyme; σ factor; promoter; operator.

8-5. It is very rare that both DNA strands in any particular region serve as a template for RNA synthesis. Describe an old observation concerning mutations in fungi from which the above result might have been predicted.

8-6. When the RNA polymerase reaction is carried out *in vitro*, it is found that at physiological ionic strength fewer regions of a particular DNA molecule are transcribed than when the ionic strength is very low. Furthermore, at low ionic strength, the regions described differ markedly from those observed to be transcribed *in vivo*. Explain.

8-7. The rate of initiation of new RNA strands is greater if the DNA template is supercoiled than if the DNA is linear or is a nicked circle. Propose an explanation.

●8-8. DNA has complementary single-stranded ends which allow circularization by means of base-pairing of these segments. DNA ligase can be used to seal the single-strand breaks; the resulting circle usually does not have any superhelical turns. If RNA polymerase is bound to the DNA before the joining of the end segments and ligation, and then the RNA polymerase is removed, the DNA circle has superhelical twists. What does this say about the binding of RNA polymerase to DNA?

○8-9. How do the modes of action of the following antibiotics differ? Actinomycin D; rifampicin; and streptolydigin.

8-10. Describe the function of each subunit of *E. coli* RNA polymerase that has been determined.

●8-11. Consider an operon having this order: repressor, operator, promoter, and cistrons *a*, *b*, and *c*. The repressor binds to the operator. When an inducer I is added, the repressor is inactivated and a single polycistronic mRNA is made. Normally, equal amounts of products A, B, and C are made; these products are needed for the utilization of I as a carbon source. A mutant is isolated which gives small colonies on an agar surface when A is the sole carbon source. Because you are interested in the proteins of so-called "leaky" mutants, you study this further. You measure activity of proteins A and B by standard biochemical tests in the presence and absence of I. You find that when I is present, the activity of A is 10 percent of normal and that of B is slightly higher than normal (normal means having the activity of the wild-type). Surprisingly, when I is absent, B activity is still found at near-normal levels, though A is absent. To study this further, you decide to try to map the mutation, perform dominance tests, check either the fingerprint or the amino acid sequence of A, and assay for C. A colleague tells you not to waste your time with this tedious work, since he can tell you the result of each test; furthermore, he is sure that the activity of the C protein will be nearly normal. What does your colleague predict?

●8-12. Assuming that you do not have available any data on the base sequence of any mRNA molecule, what information is available that suggests all promoters do not have the same base sequence?

○8-13. What chemical groups are present at the origin and terminus of a molecule of mRNA that has just been synthesized?

8-14. If the purified holoenzyme of *E. coli* RNA polymerase (the holoenzyme includes the σ subunit) is added to DNA in the presence of the four necessary nucleotide triphosphates, RNA is made. Analysis of the size of the RNA chains made shows mRNA molecules having a wide range of sizes, and the size distribution is nearly continuous. If a small amount of a cell extract (demonstrated to be free of ribonuclease activity) is added, smaller fragments of discrete size are observed. What is probably present in the extract?

○8-15. Ribosomes must of course be synthesized. Which of the following components are necessarily involved in the synthesis of a ribosome? (There are other necessary components than are included here):
  (a) DNA polymerase;
  (b) RNA polymerase;
  (c) tRNA;
  (d) ribosomes.

●8-16. The purpose of regulation of gene expression is to reduce wasted synthesis; that is, enzymes should be made only if needed. Mutants defective in regulation of a particular operon can be isolated (for example, those which have inactive repressors or inactive operators). Furthermore, we can break down the regulation by adding a *gratuitous inducer*, that is, one which is not a substrate of the induced enzyme. Consider an operon *x* which is induced by the product X and makes an enzyme which allows X to be the sole carbon source.
  (a) Suppose 10^6 wild-type and 10^6 repressorless bacteria are placed in a growth medium lacking X. The repressorless cells synthesize Xase at a level of 5 percent of the dry weight of the cell. If the bacteria are allowed to grow for 30 wild-type generation times (and the cells are diluted to prevent saturation of the culture), what will be the ratio of wild-type to repressorless cells at this time?
  (b) Consider part (a) if X is present.
  (c) Suppose the wild-type culture is grown for a very long time in the presence of a gratuitous inducer; what type of mutant might accumulate in the culture?

8-17. For each of the following diploid genotypes, indicate first whether β-galactosidase can be made, and second, whether synthesis of β-galacotosidase is inducible (I) or constitutive (C), and finally, whether or not each cell could grow with lactose as sole carbon source.
  (a) $i^+ z^- y^+ / i^- z^+ y^+$.
  (b) $i^+ z^+ y^+ / o^c i^+ z^- y^+$.
  (c) $i^+ z^- y^+ / o^c z^+ y^+$.
  (d) $i^+ z^+ y^- / i^- z^- y^+$.
  (e) $i^- z^+ y^- / i^- z^+ y^+$.
  (f) $i^- z^+ y^+ / i^+ o^c z^- y^+$.
  (g) $i^+ p^- z^+ / i^- z^-$.
  (h) $i^+ o^c z^- y^+ / i^+ z^+ y^-$.
  (i) $i^+ p^- o^c z^- y^+ / i^+ z^+ y^-$.
  (j) $i^- p^- o^c z^+ y^+ / i^- z^- y^-$.

8-18. For each of the *E. coli* diploids that follow, indicate whether the strain is inducible or constitutive, or negative for β-galactosidase *and* permease, respectively.

(a) $i^+ o^+ z^- y^+ / i^+ o^c z^+ y^+$.   (d) $i^+ o^c z^- y^- / i^- o^c z^- y^-$.

(b) $i^+ o^+ z^+ y^+ / i^- o^c z^+ y^-$.   (e) $i^- o^+ z^+ y^- / i^+ o^c z^- y^-$.

(c) $i^- o^+ z^- y^+ / i^- o^c z^+ y^+$.

8-19. Suppose you have isolated a Lac⁻ mutant and by genetic analysis have found that the cell is $z^+ y^+$; you have also found that the mutation, which you call $i*$, is in the i gene. The diploid $i* z^+ y^+ / i^- z^+ y^+$ is constructed and found to be Lac⁻; that is, $i*$ is dominant. The diploid $i* z^+ y^+ / o^c i^+ z^+ y^+$ is Lac⁺ (β-galactosidase is made). Suggest a property of the mutant repressor that gives it this phenotype. Would $i* z^+ y^+ / o^c i^+ z^- y^+$ make β-galactosidase?

8-20. Describe, in terms of mRNA synthesis, enzyme synthesis, and enzyme activity, what happens if lactose is added to a culture of a lac^+ E. coli strain previously growing in a nutrient medium lacking all sugars. Assume that the amount of lactose added is consumed after two generations of growth.

8-21. Repeat 8-20, this time for a lac^+ culture in lactose-containing medium when an amount of *glucose* is added which is consumed in one generation.

o8-22. A mutation is found in which a piece of DNA 50 base-pairs long is added between p and o of the lac operon. This has the effect of making β-galactosidase synthesis constitutive. Which of the following statements must be true?
(a) CAP-binding activity has become unimportant.
(b) The size of the repressor protein must be less than 185 Å long. (Base pairs are separated by 3.7 Å).
(c) RNA polymerase can displace a repressor from the operator.

o8-23. The action of the cyclic AMP-CAP system ensures which of the following?
(a) That cyclic AMP levels do not become too high.
(b) That enzymes are made when needed.
(c) That enzymes are not made when not needed.

8-24. A particular operon seems to be regulated by a gene called p. It has been suggested that the p-gene product is not a protein but an RNA molecule that is not translated. The principle evidence is that a protein product has not been detected. A series of p mutants are isolated and characterized. Some of these are temperature-sensitive mutants. In an effort to study these thoroughly, the mutations are transduced into strains carrying various suppressors. In two strains, one carrying an amber and one an ochre suppressor, a particular p^- mutation behaves as if it is p^+. You now know whether the p-gene product is a protein or RNA. Which is it, and how do you know?

8-25. An operon involved in utilizing a sugar, X, is regulated by a gene called b. When X is added to the cells, Xase is made; otherwise it is not. If the b gene is deleted (this is denoted by Δb), no Xase can be made. The diploid $b^+/\Delta b$ is inducible. Point mutants of b are of two types: $b1$ never makes Xase, $b2$ is constitutive. The partial diploids $b^+/b1$ and $b^+/b2$ are inducible and constitutive, respectively. What is a likely mode of action of the protein made by b?

8-26. An *E. coli* mutant is isolated that renders the cell simultaneously unable to utilize a large number of sugars, including lactose, xylose, and sorbitol. However, transduction analysis shows that each of the operons responsible for utilization of these sugars is free of mutation. What are the possible genotypes of this mutant?

o8-27. Suppose *E. coli* is growing in a growth medium containing lactose as the sole source of carbon. The genotype is $i^-z^+y^+$. Glucose is then added. Which one of the following will happen? (Only one answer is correct.)
(a) Nothing.
(b) Lactose will no longer be utilized by the cell.
(c) *lac* mRNA will no longer be made.
(d) The repressor will bind to the operator.

o8-28. A cell with genotype $i^+z^+y^+$, in a growth medium containing neither glucose nor lactose (that is, it uses some other carbon source) has how many proteins bound to the DNA comprising the *lac* operon? How many if glucose is present?

8-29. An Hfr that is *lac*$^+$ (its genotype is $i^+o^+z^+y^+$) is mated with a female that is $i^-o^+z^-y^-$). In the absence of inducer, β-galactosidase is made for a short time after the Hfr and female cells have been mixed. Explain why it is made and why only for a short time. What would happen if the female were $(i^+o^cz^-y^-)$?

8-30. What mutation of the *lac* repressor is constitutive and *trans*-dominant?

8-31. In haploid cells the i^- and o^c mutations give rise to constitutive synthesis of the structural genes of the *lac* operon. How would you use a *cis-trans* test to distinguish an i^- from an o^c mutation.

●8-32. How would you determine whether the *lac* operon is transcribed in a clockwise direction? The following article will be helpful: *J. Mol. Biol.* (1969), **40:** 145.

8-33. Although the level of enzymes made is tightly regulated, it is often found that more enzyme is made per cell if the operon is located on an F' sex factor than if it is located on a bacterial chromosome. Explain.

●8-34. In order to study the regulation of the *lac* operon in *E. coli* you perform a diploid analysis with various regulatory and structural gene mutants which you have isolated. The results of your experiments are shown in Table 8-1. The numbers represent relative activity of the enzyme β-galactosidase (the product of the z gene). For the diploids, the genotypes of the episomal and chromosomal *lac* operon are written on the left and right sides of the bar, respectively. Your results are somewhat different from what you had expected.

(a) Complete the table: at experiments 5–10, write in the expected missing numbers.

(b) Compare experiments 1 and 4. Why do the values for induced β-galactosidase differ?

(c) Compare experiments 5 and 8 with 1. Can you think of any explanation for the low values obtained in experiments 5 and 8?

(d) Examine experiments 9 and 10. Can you think of any explanations to account for these low values?

Hint: The results in experiments 5–10 are different manifestations of the same phenomenon.

TABLE 8-1

Experiment	Genotype	Observed β-galactosidase activity		Expected β-galactosidase activity	
		Induced	Uninduced	Induced	Uninduced
1	$i^+o^+z^+$	100	0.1	100	0.1
2	$i^-o^cz^+$	100	100	100	100
3*	$i^-o^cz_1^-$	0.1	0.1	0.1	0.1
4	$i^+o^+z^+/i^+o^+z^+$	200	0.1	200	0.1
5	$i^-o^cz_1^-/i^+o^+z^+$	10	0.1	____	____
6	$i^+o^cz_1^-/i^-o^+z^+$	10	0.1	____	____
7	$i^-o^+z_1^-/i^+o^cz^+$	10	100	____	____
8	$i^+o^cz^+/i^-o^+z_1^-$	10	100	____	____
9	$i^-o^+z_2^-/i^+o^cz_3^-$	40	0.1	____	____
10	$i^-o^+z_3^-/i^+o^cz_2^-$	40	0.1	____	____

* z_1^-, z_2^-, and z_3^- are three different mutant alleles in the z gene. All three are missense mutations (a wrong amino acid is inserted).

●8-35. Explain how lactose molecules first enter an uninduced $i^+z^+y^+$ cell to induce synthesis of β-galactosidase.

●8-36. Suppose that someone has just described a virus containing a linear double-stranded RNA molecule and has established that a phage-encoded RNA replicase is needed to replicate the RNA. Since this enzyme has many properties in common with *E. coli* DNA polymerase, a scientist proposes to determine whether the RNA is also synthesized discontinuously, that is, in small fragments. The experiment seems simple enough, since synthesis of both RNA and DNA will shut off immediately after the host is infected. Therefore the host cell is infected; ^{3}H-uridine is added for two seconds; RNA is isolated and sedimented through an alkaline sucrose; the contents of the centrifuge tube are fractionated; and the radioactivity in each sample is counted with a scintillation counter, merely by adding the sample to a scintillation solvent that accepts alkaline sucrose. (This is a standard procedure for making an analysis of fragments at a replication fork.) The scientist finds that all of the radioactivity is at the meniscus of the centrifuge tube. Taking this to mean that the RNA is replicated discontinuously, the scientist submits a brief communication of the finding to a periodic journal. However, the paper is rejected because of an error in the experimental method. What might it have been?

●8-37. How would you go about trying to isolate a plaque-forming, transducing phage-particle (referred to as ϕp) that carries a *lac* operon in which only the promoter region is deleted? Outline the steps you would take, in numerical sequence. The following article will be helpful: *Mol. Gen. Genet.* (1974), **129**: 201.

●8-38. We would like to have a combined genetic and biochemical analysis of chain termination for RNA equivalent to that available for chain initiation in the *lac* operon. One possible approach utilizes the large, polar insertions of DNA which are believed to cause polarity by insertion of a termination sequence for RNA synthesis into a gene. Consider a polar insertion in the *z* gene of the *lac* operon. Note that this will eliminate the activities of both the *z* and the *y* genes. How would you use such a mutation to isolate second-site mutations which might be defective in the termination system? (*Hint:* Selection for a permease-positive phenotype can be done independently of selection for a β-galactosidase-positive phenotype.) Assuming that you have such "termination" mutations, what genetic analysis would you make to decide whether they affect a termination site or the gene for a protein factor (the *rho* factor) required for termination? What *in vitro* experiments might you do to decide whether the mutations affect a site or the gene for a protein factor?

●8-39. In a hypothetical *E. coli* operon the regulator gene is closely linked to a region containing two structural genes (consider only one) and an operator. In Table 8-2, the regulator, the operator, and a structural gene are listed in correct sequence. The ability of each of the indicated gentoypes to synthesize an enzyme under induced and noninduced conditions is as shown. Which of these genes is the regulator? The operator? The structural gene? Explain.

●8-40. In an inducible bacterial operon, *a* is the only known regulatory gene (but there may be another that is unknown), *b* is the operator, and *c* and *d* are the structural genes whose wild-type alleles specify enzymes C and D and whose mutant alleles specify enzymes C⁻ and D⁻. Mutant allele a^- permits constitutive synthesis of the enzymes. This might occur in one of two ways. (1) The a^- product is defective and fails to bind to the operator. (2) The a^- product stimulates production of an internal inducer, thus eliminating the need for an external inducer.

TABLE 8-2

Genotype	Phenotype	
	Inducer absent	Inducer present
$a^- b^+ c^+$	S	S
$a^+ b^+ c^-$	S	S
$a^+ b^- c^-$	s	s
$a^+ b^- c^+/a^- b^+ c^-$	S	S
$a^+ b^+ c^+/a^- b^- c^-$	s	S
$a^+ b^+ c^-/a^- b^- c^+$	s	S
$a^- b^+ c^+/a^+ b^- c^-$	S	S

Note: S = enzyme synthesized in normal quantities; s = little or no synthesis.

The phenotypes produced by various diploid genotypes when an external inducer is present and when it is absent are shown in Table 8-3.

(a) Which alternative, (1) or (2), accounts for the phenotypes produced by the a^- allele?

(b) If alternative (2) is correct, do the a alleles act only on loci in cis positions? If not, what is the probable nature (diffusible versus nondiffusible) of the repressor? Explain.

TABLE 8-3

Genotype	Inducer present				Inducer absent			
	C	C⁻	D	D⁻	C	C⁻	D	D⁻
$a^+ b^+ c^- d^-/a^- b^+ c^+ d^+$	+	+	+	+	−	−	−	−
$a^- b^+ c^+ d^+/a^- b^+ c^- d^-$	+	+	+	+	+	+	+	+
$a^+ b^+ c^+ d^-/a^- b^+ c^- d^+$	+	+	+	+	−	−	−	−
$a^- b^+ c^- d^-/a^+ b^+ c^+ d^+$	+	+	+	+	−	−	−	−
$a^- b^+ c^- d^-/a^- b^+ c^+ d^+$	+	+	+	+	+	+	+	+

Note: + and − entries in the body of the table denote presence and absence, respectively, of the different enzyme types.

8-41. Would you expect the synthesis of repressors of catabolic operons to be regulated? Explain.

8-42. A region of the E. coli chromosome has been mapped as follows:

arg	lac	pro	tsx	pur

$$\underbrace{y \; z \; o \; p \; i}_{} \qquad\qquad \underbrace{b \; a \; o}_{}$$
$$\longleftarrow \qquad\qquad\qquad\qquad \longleftarrow$$

The arrows denote the direction of transcription and arg = arginine, lac = lactose, pro = proline, tsx = adsorption site for phage T6, pur = purine; y = lac permease, z = β-galactosidase, o = an operator, i = repressor, and b and a are genes in the purine operon. From an *E. coli* strain that is $lac^+ pro^+$ and T6-sensitive, a single-step mutant with the phenotype Lac⁻ Pro⁻ Tsx-r is isolated. This shows the following properties. (1) In the absence of lactose, lac permease is synthesized if purines are absent, but not if they are present. (*Note:* The pur operon is repressed when purines are present.) (2) If purines are present, lac permease cannot be induced by lactose. (3) No β-galactosidase activity is observed under any conditions, and this property accounts for the Lac⁻ phenotype. What kind of mutant must this be, and what does it show?

8-43. When a prophage is situated adjacent to a bacterial operon, it is often the case that enzymes from that operon are synthesized when the prophage is induced. This is called **escape synthesis.** What features of prophage and operon transcription are necessary for escape synthesis to occur?

8-44. In *E. coli*, the enzyme alkaline phosphatase (APase) is coded for by the gene *phoA*. If phosphate, a product of the reaction catalyzed by APase, is added to a cell culture, synthesis of APase is greatly inhibited in the growing cells. There are two other genes, *phoB* and *phoR*, which also seem to be involved in regulating the expression of *phoA*: mutations in *phoB* lead to reduced synthesis of APase in the presence or absence of phosphate. Mutations in *phoR* lead to constitutive synthesis of APase. (See Table 8-4.)

TABLE 8-4

Mutants	Relative level of APase when phosphate is	
	Absent	Present
$phoA^+ phoB^+ phoR^+$	100	20
$phoA^- phoB^+ phoR^+$	0	0
$phoA^+ phoB^- phoR^+$	5	1
$phoA^+ phoB^+ phoR^-$	100	100
$phoA^+ phoB^- phoR^-$	5	5

Note: Consider the *phoB⁻* to be a point missense mutation in the *phoB* gene.

We know that the *phoB* and *phoR* genes each produce a protein, so the mutations shown in Table 8-4 do not represent *cis*-acting mutations. We also know that *phoA* is being regulated by transcriptional control. Assume that no other proteins or small molecules besides the ones mentioned here are involved in regulating the *phoA* gene.

(a) Which *one* of the regulatory proteins is interacting directly with phosphate to regulate the expression of APase? Briefly explain why you concluded this.

(b) Which *one* of the regulatory proteins is interacting directly with a site on the DNA? Is this protein a positive or negative effector? Briefly explain.

(c) Explain how a complex between the *phoB* and the *phoR* proteins might function in regulating the expression of APase. Include the role of phosphate in your explanation.

●8-45. The loci of an inducible operon, in the correct order, are $a, b, c, d, e,$ $f,$ and $g.$ Loci $c, d,$ and e are structural genes for three enzymes.

Following are descriptions of properties of mutant genes $a, b, f,$ and $g.$ Study the descriptions and then formulate a model of the regulation of the operon.

Mutants of the b gene have the following properties: b^- mutants make very low levels of enzymes C, D, and E; b^+ is *trans*-dominant, and makes high levels of C, D, and E if the a gene is deleted. The activities of C, D, and E are the same regardless of the b^+ or b^- genotype if gene f is deleted. Amber mutants of b exist.

Mutants of a fall into three classes: (1) One class results in constitutive expression of C, D, and E. Such mutants are uninducible in strains which are also $b^-.$ The a^- mutations in this class are recessive. (2) The second class also results in constitutive expression of the operon in strains which are $b^-.$ Unlike class (1), mutations of this class are dominant. (3) The third class of a^- mutants lack operon function both in the presence and absence of inducer. This is true if the strain is also $b^-.$ At very high inducer-concentrations, however, some induction can be shown to take place.

Mutants f^- in strains in which a is deleted and which are also b^- have low constitutivity (10–25 percent of that induced in the wild-type) and are *cis*-dominant.

Mutants g^- are *cis*-dominant. If a strain is both g^- and $f^-,$ the operon is expressed constitutively at high levels.

None of the mutant types affect any other operon.

Now, make a model that is consistent with these data. *Hint:* The effects of b and a should be analyzed separately before stating a comprehensive model.

8-46. As an aid to an analysis of repressor-polymerase interactions, you isolate a large number of *cis*-dominant constitutive mutations. Draw an imaginary genetic map of ten mutations and state their phenotype under each of the following assumptions:

(a) Operator and promoter sites are identical.

(b) Operator and promoter sites overlap.

(c) Operator and promoter sites are completely independent.

8-47. The regulation of an operon responsible for synthesis of X is dependent on a repressor, a promoter, and an operator. In the presence of X, the system is turned off; an interaction between X and the repressor forms a complex which can bind to the operator.

(a) What kinds of mutations might occur in the repressor? Describe their phenotype (in terms of whether the operon is on or off).

(b) Describe the phenotype of a partial diploid with a wild-type and each mutant gene.

●8-48. In analyzing regulation of an inducible galactose operon you find that there seem to be two regulatory genes (*galR* and *galS*) unlinked to the structural genes of the operon and to each other. Mutations in either gene (including nonsense-type) give a constitutive phenotype, as shown in Table 8-5. The constitutive mutations in either gene are recessive to wild-type in both *cis*- and *trans*- configurations.

(a) Propose three models of regulation for this operon.

(b) What additional types of mutations might you seek to determine which of the three models is more likely?

TABLE 8-5

| | Level of gal enzymes | |
Mutation	Uninduced	Induced
$r^+ s^+$	1	100
$r^- s^+$	20	100
$r^+ s^-$	5	100
$r^- s^-$	100	100

8-49. You wish to distinguish between a "competitive" mechanism for repression and one which allows binding of both RNA polymerase and repressor. Assume that you can use restriction enzymes to prepare DNA (labeled or unlabeled) with a single promoter and operator site. How might you use this DNA to try to distinguish between the two possible mechanisms? (Assume also that you can prepare pure RNA polymerase, repressor, or any other DNA molecule that you wish to use.)

●8-50. The pathway for biosynthesis of biotin is given below. Arrows represent the enzymes that catalyze each step and the letters above the arrows represent the genes that code for these enzymes:

$$\overset{c}{X} \rightarrow \overset{f}{PCA} \rightarrow \overset{a}{KAPA} \rightarrow \overset{d}{DPA} \rightarrow \overset{b}{DTB} \rightarrow \text{Biotin.}$$

The five genes a, b, c, d, and f are clustered on the E. coli map and form the *bio* operon. A detailed map of the *bio* operon looks like this:

$$\underline{ \quad a \quad o_L \quad p \quad o_R \quad b \quad f \quad c \quad d}$$
$$gal \quad att\lambda \qquad\qquad\quad bio$$

In this map $att\lambda$ means the chromosomal attachment site for prophage λ, o_L and o_R are the leftward and rightward operators of the *bio* operon, and p is a promoter which serves all of the genes.

(a) You can make λ*bio*-specialized transducing-phage particles and use them to make mRNA molecules *in vitro*. By hybridization of the mRNA molecules with $\phi80bio$-specialized transducing-phage particles, you can purify *bio*-operon mRNA molecules because the λ- and $\phi80$-DNA molecules are not homologous. Do you expect the *bio* mRNA to hybridize with the following DNA strands (H and L indicate the heavy and light strands, respectively, of the phage DNA; the heavy strand is the coding strand for λ late-stage mRNA):

λbio, H; λbio, L; λ, H; λ, L.

(b) You have isolated a polar amber mutation in the f gene. Predict the derepressed levels of enzymes for the five *bio* genes. What effect would a polar mutant in gene a have on genes $b, f, c,$ and d?

(c) A mutant of E. coli has been isolated that never produces the *bio* enzymes; that is, they are absent even when the mutant is grown on low amounts of biotin. This is known to be a mutation in a gene that codes for a repressor of the *bio* operon. Do you expect this mutant to be dominant or recessive to wild-type? In *trans* or *cis*?

(d) A bacterium related to E. coli requires biotin since it is unable to make its own; therefore biotin must be added to the medium for this bacterium to grow. However, this bacterium is known to possess the *bio* genes because λ*bio*-specialized transducing-phage particles prepared from the bacteria can lysogenize E. coli cells in which the *bio* operon has been deleted, and the resultant lysogen is bio^+ (that is, it does not require the addition of biotin). Also the bacteria can mutate at very low frequency to gain the ability to grow in the absence of biotin; these mutants produce the *bio* enzymes constitutively. Suggest an explanation.

8-51. Another useful procedure for analyzing mRNA is **hybridization competition.** In this procedure a small and constant amount of radioactive mRNA is mixed with larger amounts of nonradioactive mRNA and with a fixed amount of single-stranded DNA that is bound to a nitrocellulose filter. The various mRNA species then form hybrids with the DNA. If the nonradioactive mRNA contains sequences also present in the radioactive mRNA, then at very high concentrations of mRNA, when the amount of DNA is less than that of the mRNA, the amount of radioactive mRNA that hybridizes will be small compared to the amount of nonradioactive mRNA. However, if there are no common sequences, there will be no competition with the radioactive mRNA during hybridization. Thus the temporal sequence of mRNA synthesis can be analyzed by noting whether nonradioactive mRNA isolated at a time t_1 diminishes the ability of a radioactive mRNA molecule isolated at a time t_2 to hybridize. For example, if the amount at t_1 is greater than the amount at t_2, and the ability to hybridize does not diminish, then the mRNA made at t_2 is no longer being made at t_1.

Suppose that nonradioactive mRNA has been isolated from phage-infected cells at two, three, four, and six minutes after infection. This is then used in hybridization-competition experiments and the data in Figure 8-1 are obtained.

What can you say about the stability of the two-minute mRNA and the timing of its synthesis?

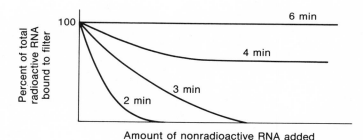

FIGURE 8-1

8-52. Phage T7 DNA (molecular weight = 26×10^6) is transcribed in two stages. *E. coli* RNA polymerase (RNAP) copies the early genes, one of which codes for a phage T7 RNA polymerase. The T7 polymerase then transcribes T7 late genes. Figure 8-2 shows a transcription map of T7 DNA (numbers refer to molecular weights of the mRNA species in units of 10^6 daltons).

Phage T3 is related to phage T7, has DNA of the same size, and also induces its own RNA polymerase. T3 mutants which make a temperature-sensitive RNA polymerase do not synthesize the major T3 head-protein at 42° C.

From the following data, construct a transcription map of T3 DNA: (1) *E. coli* RNAP plus T3 DNA yields a homogeneous product of 13×10^6 daltons. (2) In the presence of termination factor *rho*, transcripts of 2×10^6 daltons are produced by *E. coli* RNA polymerase. (3) T3 RNAP copies T3 DNA to yield four discrete-sized transcripts having molecular weights of 2, 5, 8, and 11×10^6.

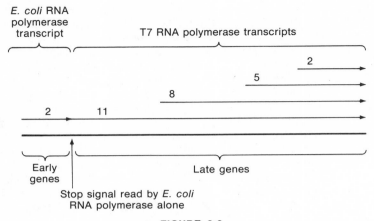

FIGURE 8-2

These transcripts were analyzed by the hybridization-competition procedure (see problem 8-51), with the results shown in Figure 8-3. What is the difference between the T3 and T7 transcription maps?

8-53. How would you show that the T7 phage-induced RNA polymerase can also transcribe the early phage T7 genes *in vivo?*

8-54. How would you go about demonstrating whether or not the late *in vivo* T7 transcripts, which are subsequently cleaved to units of smaller molecular weight, are polycistronic?

8-55. A genetic engineer has constructed a phage DNA molecule having the sequence of signals and genes shown in Figure 8-4. Transcription is from right to left. Site B = structural gene for T7 RNA polymerase; N = gene N of phage λ; p_L = normal promoter in the early leftward transcription unit of phage λ; p_{EFG} = normal promoter used by E. coli RNA polymerase in reading genes E, F, and G; $p_{T7-RNAP}$ = promoter used by T7 phage-induced RNA polymerase; t_L = normal terminator in the early leftward transcription unit of phage λ; solid square = a normal *rho*-dependent terminator; $t_{T7-late}$ = a terminator for T7 RNA polymerase; $t_{T7-20\text{ percent}}$ = the kind of terminator for E. coli RNA polymerase found at 20 percent of the distance between the left and right ends of T7 phage DNA; solid circle = a polar amber mutation.

(a) Indicate which regions are transcribed *in vivo* by E. coli RNA polymerase (EP) and T7 RNA polymerase (TP) in each of the seven situations below (that is, trace the map on a sheet of paper and then draw arrows, labeled EP or TP, below the genetic map, using the convention that the tip of each arrow denotes the site of actual termination and the base of the arrow indicates the site where tran-

Percent initial hybridization to T3 DNA

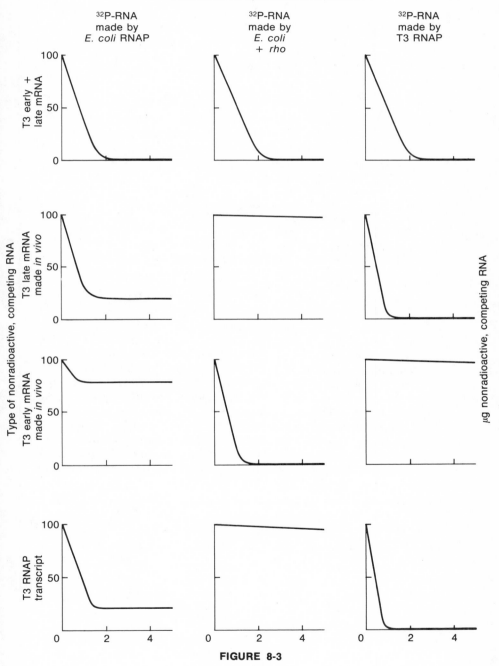

FIGURE 8-3

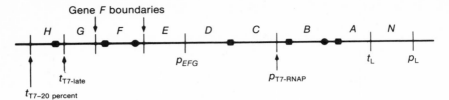

FIGURE 8-4

scription is initiated). Note that gene *B* provides the only source of T7 RNA polymerase.

Situations:

1. No amber mutations anywhere.
2. Amber mutation in gene *F*, as indicated.
3. Amber mutation in gene *N*.
4. Amber mutations in genes *F* and *N*.
5. In a host with a temperature-sensitive mutation in *rho* and at a nonpermissive temperature. Also, there is an amber mutation in gene *F*.
6. Amber mutations in genes *B* and *N*.
7. *N* gene is from *imm*21, but p_L is from λ.

(b) What is the strength of expression of genes *B*, *D*, *F*, and *G*, in each situation in (a)? Assume that transcription from each of several promoters contributes an expression efficiency of 1.0.

8-56. Messenger RNA is usually measured by various hybridization techniques—that is, through the formation of a DNA-RNA hybrid by incubation of the mRNA with single-stranded DNA. In one technique, single-stranded DNA is adsorbed to a nitrocellulose filter, radioactive RNA is added, and then the mixture is incubated under conditions leading to hybridization. Free RNA does not bind to the filter. Therefore, the amount of radioactivity bound to a washed filter is a measure of the amount of RNA hybridized. Radioactive mRNA is isolated from a phage-infected cell one minute after infection. This is hybridized with three different DNA molecules separately adsorbed to filters. The DNA molecules are

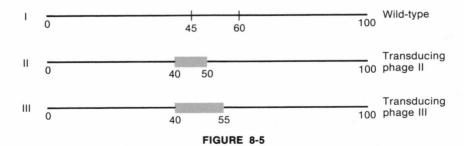

FIGURE 8-5

shown in Figure 8-5. Numbers refer to distance from the left end of the DNA, on a scale of 0–100. Wild-type (I) phage was used in the infection. The shaded regions in II and III represent bacterial DNA. Following hybridization with each type of DNA on two separate filters, one filter of each type is incubated with a ribonuclease (+RNase) that digests single-stranded RNA but not hybridized RNA; the other filter is untreated (−RNase). The data in Table 8-6 are obtained. From which region of the wild-type DNA was the mRNA transcribed?

TABLE 8-6

DNA on filter	cpm on filter	
	−RNase	+RNase
I	1,250	1,245
II	1,260	820
III	1,242	418

9

The Genetic Code

Introduction

The sequence of bases in DNA determines the order of amino acids in each protein. The relation between these two sequences is known as the genetic code.

PROPERTIES OF THE CODE

The genetic code is a three letter alphabet—that is, each group of three bases corresponds either to an amino acid or to a start or stop signal (see Appendix A, page 237). The code is nonoverlapping, which means that a single base can be a part of only one codon at a time. For example, consider the sequence of bases AGUGUCAGA. In a non-overlapping triplet code, this sequence is read as three codons, AGU, GUC, and AGA. If the code were overlapping, it would be read as seven codons, AGU, GUG, UGU, GUC, UCA, CAG, and AGA. An overlapping code contains more information in a given sequence of DNA but has two main defects. (1) A single base change (that is, a mutation) would always alter three amino acids so that mutations would almost always be deleterious. (2) The number of possible amino acid sequences would be considerably reduced. For example, if a triplet code were overlapping, each sequence of four bases would code for two amino acids (a dipeptide). Since there are $4^4 = 256$ different tetranucleotides, this would be the maximum number of possible dipeptide sequences in proteins. As the code is nonoverlapping, each of the 20 amino acids can be followed by any other (there can be $20^2 = 400$ possible dipeptides). Even before the code was worked out by chemical means, it was known to be nonoverlapping for two reasons. First it was found that there is rarely a change of more than one amino acid in a mutant protein; second, more than 256 dipeptide sequences were already known to occur in proteins.

There is no obvious way to predict from chemical or evolutionary principles which codon should correspond to a particular amino acid. However, the codon sequences are not randomly related to the amino acids, as the following features of the code indicate:

1. The code is redundant. That is, except for tryptophan and methionine, each of which has only one codon, every amino acid is coded for by at least two codons; nine amino acids have two, one has three, five have four, and three have six codons. There are three chain-termination codons. This redundancy minimizes the deleterious effect of mutations.

2. The third base of a codon frequently carries little or no information. For eight amino acids a change in the third base of a codon does not alter the amino acid specified. Again, this minimizes the effect of mutation. (See the Introduction to Chapter 10 for information about this point.)

3. The second base determines the polarity of the amino acid specified. In each case, if the amino acid has a polar side chain, the second base is a purine; if the side chain is nonpolar, the second base is a pyrimidine. This has the effect that transition mutations in this position usually have little or no effect on protein structure. Only transversions can induce major changes in protein structure.

ELUCIDATION OF THE CODE

The triplet nature of the code was first shown by genetic analysis. Francis Crick and his associates studied acridine-induced frameshift mutations in a nonessential region of the *rII* gene of *E. coli* phage T4. This region was known to be nonessential in the sense that mutations in this region were rarely found, and the few that were found were very leaky; in fact, the region could even be totally deleted. However, frameshift mutations in this region could be easily induced by acridines. These mutations could not be reverted except by exposure to mutagens inducing frameshift mutations. These mutations fell into two classes, +

and −, according to the criterion that if two mutations were combined in a single phage by genetic recombination, a functional *rII* protein resulted only if one mutation came from the + class and the other from the − class. Since the mutations were all frameshifts, one class was assumed to consist of a single base addition and the other a single base deletion. Thus, the combination of a + with a − restored the correct reading-frame before the essential part of the protein was reached. Proof of the triplet nature of the code came from the observation that a combination of two or four + mutations or the same number of − mutations would not yield the wild-type phenotype, but a combination of three + or three − mutations would. That is, adding three bases or deleting three bases does not alter the reading-frame. It should be noted that these experiments were possible only because it was possible to make numerous amino acid substitutions, additions, and deletions in a nonessential part of the *rII* protein.

The chemical nature of the code was first worked out by Marshall Nirenberg, Severo Ochoa, Gobind Khorana and their co-workers following the accidental discovery by Marshall Nirenberg, in 1961, that the addition of polyU to a system for synthesizing protein *in vitro* resulted in the formation of an insoluble polypeptide, polyphenylalanine. Initially, the composition of each codon was determined in an *in vitro* protein-synthetic system using synthetic polyribonucleotides as mRNA. Mixed polymers with known base ratios were prepared; the ratios of the number of triplets having each composition were calculated (for example, in polyUG, the relative frequency of triplets of three U, two U and one G, two G and one U, and three G); and the molar ratios of the amino acids incorporated were measured and compared to the calculated codon ratios. This technique was refined to give the base sequence of the codons by using long homopolymers terminated by a small number of bases of a second type. For instance, if polyU is terminated on the average by a single G, in order of decreasing frequency, the codons are UUU, UUG, UGG, and GGG. Later it was shown that a synthetic trinucleotide of known sequence would stimulate the binding of a specific tRNA molecule to ribosomes; since each tRNA species can be labeled by the amino acid it binds, all of the codons could be identified. The final element in codon determination came when synthetic polyribonucleotides of defined sequence, for example, GACGAC-GACGACGAC, were used to direct protein synthesis and the sequence of amino acids in the protein thus synthesized could be determined.

The above work clearly defined the codon assignments for the crude protein synthetic systems available in the laboratory. That these are the correct codons *in vivo* was confirmed a few years later when the complete sequences of amino acids in the coat protein and the bases of the coat-protein gene of *E. coli* RNA phage MS2 were determined.

REFERENCES

Cold Spring Harbor Laboratory. 1966. *The Genetic Code. Cold Spring Harbor Symposia in Quantitative Biology, 31.* This contains numerous classic papers published during an exciting time.

Crick, F. H. C. 1968. "The origin of the genetic code." *J. Mol. Biol.,* 38: 367–379.

Lehninger, A. L. 1975. *Biochemistry.* Second Edition. New York. Worth Publishers, Inc. Chapter 31.

Stent, G., and R. Calendar. 1978. *Molecular Genetics.* San Francisco. W. H. Freeman and Company. Chapter 16. A good account of the history of the elucidation of the genetic code.

Stryer, L. 1975. *Biochemistry.* San Francisco. W. H. Freeman and Company. Chapter 25.

Watson, J. D. 1976. *Molecular Biology of the Gene.* Third Edition. Menlo Park, Calif. W. A. Benjamin. Chapter 13.

Woese, C. R. 1972. *The Genetic Code.* New York. Harper & Row.

Ycas, M. 1969. *The Biological Code.* Vol. 12 in Frontiers of Biology Series. New York. Elsevier-North Holland.

Problems

Refer to the genetic code in Appendix A, page 237, for many of the following problems.

○9-1. A DNA molecule has the structure

W	T A C G G G A A T T A G A G T C G C A G G A T C
C	A T G C C C T T A A T C T C A G C G T C C T A G

The W strand (Watson strand) is the coding strand and is transcribed from left to right. What is the amino acid sequence of the protein coded for?

○9-2. Which of the following amino acid changes can result from a single base change?

(a) Methionine to arginine.   (d) Proline to alanine.

(b) Histidine to glutamic acid.   (e) Tyrosine to valine.

(c) Glycine to alanine.

○9-3. Which amino acids can replace arginine by a single base-pair change?

9-4. The amber codon UAG does not correspond to any amino acid. Some strains carry suppressors which are tRNA molecules mutated in the anticodon so that an amino acid can be put in at a UAG site. Assuming that the anticodon has only a single altered base, which amino acids could be inserted at a UAG site? At a UAA site?

9-5. Suppose that in another world the code was a doublet code having four bases. What is the maximum number of amino acids that could be in proteins?

9-6. There are several arginine codons. Suppose you had a protein that contained only three arginines (Arg-1, Arg-2, Arg-3). In a particular mutant, Arg-1 is replaced by glycine. In another mutant, Arg-2 is replaced by methionine. In still another mutant, Arg-3 is replaced by isoleucine. Suppose several hundred other mutants at various sites are isolated. Which other amino acids would you expect to find replacing Arg-1, Arg-2, and Arg-3, assuming only single-base changes?

9-7. You have isolated an acridine-induced mutant of a particular protein and have purified the mutant protein. If the wild-type and mutant proteins are cleaved with trypsin and the peptides are examined, it is found that only one peptide is altered.

(a) Where is this peptide located in the protein? The peptide from the wild-type protein is Leu-Met-Ser-Val-Glu; that from the mutant is Leu-Thr-Glu-Arg.

(b) How many extra bases have been added to the DNA by growth in the presence of acridine orange? Identify and locate the new base or bases in the base sequence.

9-8. Which base-substitution mutagens can cause a reversion to the original amino acid by replication of a mutant DNA in the presence of the same mutagen that induced the original mutation?

○9-9. Suppose that you are making use of the alternating copolymer GUGUGUGUGU. . . as an mRNA in an *in vitro* protein-synthesizing system. Assuming that an AUG start codon is not needed in the *in vitro* system, what peptides are made by this mRNA?

9-10. Suppose you are synthesizing a random copolymer of G and U using equimolar quantities of G and U nucleotides. If this polymer is used as mRNA, which amino acids can be incorporated into protein and at what relative frequencies?

9-11. For an *in vitro* protein biosynthesis experiment to study the genetic code, a synthetic mRNA species was synthesized with poly-nucleotide phosphorylase, using adenosine diphosphate (ADP) and cytidine diphosphate (CDP) in the ratio 5 : 1. The base sequence produced in this synthesis is random. The following amino acids were incorporated into polypeptide material in the following relative amounts: lysine 100, asparagine 20, glutamine 20, threonine 24, proline 4.8, histidine 4. Using this information, identify possible codons for the above amino acids.

9-12. Which amino acids can be incorporated into a protein if the following repeating polymers are used as mRNA molecules? Assume a special start codon is not needed *in vitro*.

(a) CGACGACGACGA. . .

(b) AUGAUGAUGAUG. . .

(c) AUAAUAAUAAUA. . .

●9-13. The first codon which was identified was UUU for phenylalanine. The insolubility of polyphenylalanine allowed several other codons to be identified by a simple trick. Each nucleotide X was separately coupled to the 3′-OH terminus of polyuridylic acid (polyU). The coupling was heterogeneous in that some polyU molecules terminated with one X moiety, some with two, some with three, etc. However, the addition of a single X was more probable than two, and two more probable than three, etc.

(a) Which amino acids and in what ratio would be at the terminus of polyphenylalanine when each of the four bases was separately coupled to polyU?

(b) If each of the sixteen dinucleotides could be coupled to polyU, could the entire code be worked out? Explain.

9-14. A triplet code is nonoverlapping if the sequence ABCDADCBA can be read only as ABC-DAD-CBA and not as ABC-BCD-CDA-DAD-ADC. . . Could the method of problem 9-12 be used to determine whether the code is overlapping? What in fact is the best evidence against an overlapping code?

9-15. Considering both the code as we know it and the wobble hypothesis (see Chapter 10), what is the minimum number of tRNA molecules needed to recognize the 61 codons corresponding to amino acids?

9-16. State various hypotheses to explain the fact that the start codon AUG frequently occurs within a protein yet is not used as a start. What is the only circumstance under which a stop codon can occur within a protein yet not cause termination? (Do not consider the possibility of a suppressor.)

9-17. Human hemoglobin is a major oxygen-transporting protein and is the sole protein in red blood cells. Red blood cells have no nucleus or DNA; thus, hemoglobin is synthesized from a stable mRNA, whose 3′-terminal nucleotide sequence is

$$5'\text{-pAAGUAUCACUAAGC-}3'\text{-OH.}$$

The carboxyl-terminal dipeptide of hemoglobin is NH_2-Tyr-His-COOH. A number of mutant hemoglobins from patients with genetic diseases exhibit elongated polypeptide chains. One such mutant hemoglobin ends in NH_2-Tyr-His-Leu-Ser-COOH.

What is the mutation responsible for this elongated hemoglobin?

Hint: Only one base is involved. There are no known nonsense suppressor mutations in human cells, so the mRNA for hemoglobin must itself be altered. Assume that protein synthesis can terminate at the 3' end of a molecule of mRNA as well as at a termination signal before the 3' end.

9-18. The diagram below represents a piece of DNA from the middle of an active wild-type gene. The hyphens indicate the reading frame of the corresponding mRNA. X and Y are two unknown bases; the prime indicates the complementary base.

> 5' -X-ATA-TAG-GGG-GCA-Y-   3'   Strand 1
> 3' -X'-TAT-ATC-CCC-CGT-Y'- 5'   Strand 2

(a) What is the sequence of amino acids in the indicated four residues? Label the amino and the carboxyl termini.

(b) The change of a single base pair will give rise to a nonsense codon from the above sequence. Give the possible base-pair changes that will give a nonsense codon.

(c) In a certain mutant you find the amino acid sequence -Cys-Pro-Tyr-Met-. Assuming the occurrence of only a simple mutational event (that is, one not involving several base pairs), how does the mutant DNA sequence differ from that of the wild-type? What type of mutation is this?

(d) What must base X be, and why?

9-19. For this problem, refer to Appendix C, page 239.

(a) At position 33 in the tobacco mosaic virus coat protein, one of the amino acid substitutions occurring by mutation appears to result from two RNA base substitutions. Which amino acid is it and what are the two base substitutions?

(b) One of the changes at position 33 arose by a single transition mutation and one by a single transversion. Which is which? Indicate the new amino acid whose presence is the result of the transition and the transversion.

(c) At position 107 in the tobacco mosaic virus (TMV) coat protein, what RNA base substitution accounts for the thrice-observed mutation?

9-20. Suppose genetic recombination were a frequent, demonstrable phenomenon in TMV (which it is not). Now suppose the mutant strain with lysine at position 46 were crossed with one of the mutant strains which has glycine at position 46. Among the progeny virus produced in the cross, suppose four kinds of virus were found. By far the most frequent two of them are the parental types. What are the amino acids at position 46 in the other two types?

10

Translation and Protein Synthesis

Introduction

The information contained in the base sequence of mRNA is converted to the amino acid sequence of a protein by the process of *translation.*

TRANSFER RNA

A DNA molecule does not serve directly as a template for protein synthesis. One reason for this is probably that a single amino acid is unable to recognize and bind to three bases simultaneously. Also, if a DNA molecule were to be used for the synthesis of every protein molecule, by having to engage so frequently in chemical interactions, the DNA might be unnecessarily susceptible to attack by nucleases and other deleterious agents. Thus, *messenger RNA* (mRNA) has evolved as an intermediate that takes the information from the DNA molecule, and mRNA molecules can be used many times in protein synthesis. However, because a codon in an mRNA molecule is also unable to bind an amino acid, another molecule, *transfer RNA* (tRNA), has evolved that can bring a particular amino acid to a codon. Transfer RNA itself cannot distinguish one amino acid from another; the recognition of different amino acids is done by *aminoacyl synthetases,* enzymes which recognize both a specific amino acid and a specific base sequence in each tRNA molecule (the *recognition loop)* and which covalently link the amino acid to the amino acid binding-terminus of the tRNA molecule having the anticodon corresponding to that amino acid. Thus, the codon sequence in mRNA is translated into an amino acid sequence by a pairing of aminoacyl tRNA anticodons to the codons in mRNA. This alignment does not take place on a free mRNA molecule but only when the mRNA is bound to a ribosome, a complex particle consisting of several *ribosomal RNA* (rRNA) molecules, a large number of *ribosomal proteins,* and numerous enzymes involved in protein synthesis. The ribosome is the unit on which peptide bond formation occurs. Details of peptide bond formation and the movement of the growing polypeptide chain and the mRNA along the ribosome can be found in the references at the end of this introduction.

A unique tRNA molecule is not needed for each codon, a fact predicted by Francis Crick in his *wobble hypothesis.* The term "wobble" refers to the fact that in codon-anticodon pairing, the base pairing of the first and second bases of the codon follows the strict rule adenine to uracil and guanine to cytosine, whereas there is some steric freedom (or "wobble") in the third position; there, other weaker base pairs can form. For instance, in the third position of a codon, A can pair to U or to inosine (I), G to C or U, C to G or I, and U to A, G, or I. This is consistent with the finding (see Chapter 9) that the third base in the codon frequently is unimportant in specifying an amino acid and significantly reduces the number of tRNA types which are necessary. It should be noted that part (but not all) of the redundancy in the code is due to wobble.

N-FORMYLMETHIONINE AND INITIATION OF SYNTHESIS

The initiation of synthesis of a protein chain requires a particular tRNA molecule, a sequence of bases longer than that in the codon corresponding to this tRNA molecule, and an enzyme for forming an initiator amino acid. In bacteria, protein synthesis always starts with a modified amino acid *N-formylmethionine (fMet).* This molecule has a blocked amino group (and, properly speaking, is not an amino acid); a peptide bond can be formed only with its carboxyl group. In this way N-formylmethionine defines the *direction* of protein synthesis. An fMet-tRNA molecule is used for initiation. It differs from methionyl-tRNA, which is used only to specify methionine *within* a protein chain (although a common methionyl acyl synthetase puts methionine onto both the initiator tRNA and methionyl-tRNA). Formylation of the methionine occurs enzymatically only after attachment to the tRNA.

The codon AUG is required both for initiation and to specify internal methionine molecules. However, the fMet-tRNA does not bind to AUG sequences coding for internal methionines, and Met-tRNA usually fails to bind to AUG triplets signifying initiation. Presumably the sequence adjacent to the initiating AUG-codon is required both for recognition of fMet-tRNA and inhibition of binding of Met-tRNA. The molecular basis of this discrimination is not yet known.

N-Formylmethionine is always found at the amino terminus of proteins during synthesis of the protein chain. However, it is rarely found in completed proteins because shortly after termination of synthesis the formyl group is removed from the completed protein. This is done in two ways—either by removal of the entire amino acid or by removal of the formyl group. In *E. coli* 50 percent of the different protein types terminate with methionine so that both mechanisms occur with roughly equal frequency.

TERMINATION OF PROTEIN SYNTHESIS

Transfer RNA molecules corresponding to the codons UGA, UAA, and UAG do not exist. These chain-termination codons are instead recognized by proteins called *release factors*. There are two such factors, one recognizing both UAA and UAG, the other recognizing both UAA and UGA. In a way which is not understood, the release factors activate an enzyme which cleaves the polypeptide chain from the tRNA that is bound to the mRNA. This enzyme, *peptidyl transferase*, normally catalyzes the formation of the peptide bond through a reaction of the carboxyl group attached to the tRNA at the growing end of the protein with the amino group of the amino acid to be added. The release factors modify the specificity of peptidyl transferase so that the carboxyl group reacts instead with water; this results in separation of the protein chain from the tRNA.

ALTERATIONS OF PROTEINS AFTER TERMINATION

As stated above, the formyl group or formylmethionine itself is removed from the amino terminus of completed chains. This is an example of *post-translational modification* of proteins. Many other examples are known, a few of which are the following:

1. Removal of several N-terminal amino acids.
2. Chemical alteration of particular amino acids, such as (a) the conversion of proline and lysine to hydroxyproline and hydroxylysine respectively in the protein collagen and (b) the acetylation of histones.

3. Addition of carbohydrates to asparagine, serine, and threonine in glycoproteins.

4. Addition of lipids in lipoproteins.

5. Formation of disulfide bonds between cysteine.

6. Internal cleavages such as in the conversion of prothrombin to thrombin or proinsulin to insulin.

REFERENCES

Clark, B. F. C., and K. A. Marker. 1968. "How proteins start." *Scientific American,* **January:** 36–42.

Cold Spring Harbor Laboratory. 1969. *Mechanisms of Protein Synthesis. Cold Spring Harbor Symposia in Quantitative Biology, 34.*

Ehrenstein, G. von. 1970. "Transfer RNA and amino acid activation." *In* Anfinsen, C. B., *Aspects of Protein Biosynthesis.* New York. Academic Press. pp. 139–214.

Ingram, V. M. 1972. *Biosynthesis of Macromolecules.* Menlo Park, Calif. W. A. Benjamin. Chapter 5.

Kurland, C. G. 1972. "Structure and function of the bacterial ribosome." *Ann. Rev. Biochem.,* **41:** 377–408.

Lehninger, A. L. 1975. Biochemistry. Second Edition. New York. Worth Publishers, Inc. Chapter 30.

Losick, R. 1972. "In vitro transcription." *Ann. Rev. Biochem.,* **41:** 409–446.

Lucas-Lenard, J., and F. Lipmann. 1971. "Protein biosynthesis." *Ann. Rev. Biochem.,* **40:** 409–448.

Stent, G., and R. Calendar. 1978. *Molecular Genetics.* San Francisco. W. H. Freeman and Company. Chapter 17.

Stryer, L. 1975. *Biochemistry.* San Francisco. W. H. Freeman and Company. Chapter 26.

Watson, J. D. 1976. *Molecular Biology of the Gene.* Third Edition. Menlo Park, Calif. W. A. Benjamin. Chapter 12.

Problems

○**10-1.** A DNA molecule has the following base pair sequence:

TACGTAATAACATCACCACGGATT	Watson strand
ATGCATTATTGTAGTGGTGCCTAA	Crick strand

↑

For this gene the Watson strand is transcribed into mRNA and amino acids are assembled from left to right.

(a) What is the sequence of amino acids in the protein?

(b) In a mutant, the AT pair indicated by the arrow is reversed to TA (the A is then in the Crick strand). What is the amino acid sequence of the protein made from the mutant DNA?

(c) The tRNA molecules corresponding to which *codons* can mutate by *one* base change to allow the complete protein to be made?

(d) Write the amino acid sequences of all possible proteins which could be made using the tRNA molecules of part (c).

○10-2. The aminoacyl synthetase must

(a) Recognize the codon.

(b) Recognize the anticodon.

(c) Recognize the recognition loop.

(d) Be able to distinguish one amino acid from another.

○10-3. A tRNA molecule must be able to

(a) Recognize a codon.

(b) Recognize an anticodon.

(c) Distinguish one amino acid from another.

(d) Recognize DNA molecules.

○10-4. Which of the following are steps in protein synthesis?

(a) Binding of tRNA to a 30*s* ribosome.

(b) Binding of tRNA to a 70*s* ribosome.

(c) Coupling of an amino acid to a ribosome by an aminoacyl synthetase.

(d) Separation of the 70*s* ribosome to form a 30*s* and 50*s* ribosome.

○10-5. Which of the following are true?

(a) Transfer RNA is needed because amino acids cannot stick to mRNA.

(b) Transfer RNA molecules are much smaller than mRNA molecules.

(c) Transfer RNA molecules are synthesized without the need for intermediary mRNA.

(d) Transfer RNA molecules bind amino acids without the need of any enzyme.

(e) Some stop codons are recognized by a *rho* factor.

○10-6. The role of tRNA is

(a) To attach the amino acids to one another.

(b) To bring the amino acids to the correct position with respect to one another.

(c) To increase the effective concentration of amino acids.

(d) To attach the mRNA to the ribosome.

10-7. Suppose a chemical were introduced to a cell so that immediately after isoleucine was attached to any isoleucyl-tRNA the isoleucine is converted to valine. The result would be that in subsequent synthesis of proteins

(a) All isoleucine positions would contain valine instead.

(b) There would be no protein synthesis.

(c) There would be premature chain termination.

(d) Some isoleucine positions would be filled with valine.

○10-8. Approximately what molecular weight of DNA would correspond to all of the tRNA of *E. coli?* What fraction of the total *E. coli* chromosome would that be?

○10-9. Approximately what molecular weight of DNA would be in a gene which codes for a protein having a molecular weight of 50,000?

10-10. Suppose polyribosomes did not exist and only a single ribosome could bind to a mRNA at any moment. Compared to a system having polyribosomes, will the number of protein molecules synthesized per unit time be greater, less, or the same? Explain.

○10-11. Chain termination normally occurs because

(a) The tRNA corresponding to a chain-termination triplet cannot bind an amino acid.

(b) There is no tRNA with an anticodon corresponding to a chain-termination triplet.

(c) Messenger RNA synthesis stops at a chain-termination triplet.

○10-12. A viral RNA has the following base sequence near its 5′ end.

5′ CGAAGAUGCCCUUUGUAGCCAG. . .

(a) What are the first five amino acids of the protein?

(b) Does the UAG codon in positions 16–18 cause termination? Explain.

○10-13. Will the amino acid coded for by the codon in the 5′ direction and immediately adjacent to UAG have a free amino or free carboxyl group?

○10-14. In an *in vitro* protein-synthesizing system that requires an AUG in the mRNA, all proteins start with formylmethionine. Suppose the mRNA has the sequence AUGAAAAAAAAA. . . .

(a) What is the amino acid sequence of the protein made?

(b) Suppose an antibiotic is added which inhibits translocation. What will be the product, if any, of the *in vitro* synthesis?

10-15. Which steps in protein synthesis are inhibited by the antibiotics chloromycetin, puromycin, tetracycline, fusidic acid, and streptomycin?

10-16. Explain the function of an initiation factor, an elongation factor, and a termination factor.

10-17. Which processes in protein synthesis require hydrolysis of GTP?

10-18. In a polycistronic mRNA, it is often the case that each protein encoded in the mRNA is not synthesized to the same extent.

(a) One common phenomenon is that the amount of each protein synthesized decreases continually from one end of the mRNA to the other. This is called *polarity.* What is the current explanation for polarity?

(b) It is equally frequent that there is no obvious pattern to the amounts of the different proteins synthesized. When this occurs in phage mRNA molecules, the relative amounts made are usually exactly the ratios needed for economy (that is, there is no waste). This is called *translational control*. What are possible mechanisms for this control?

10-19. Suppose you have isolated a temperature-sensitive mutant of *E. coli*. At 42° C it is defective in protein synthesis. A culture growing at 30° C is warmed to 42°; at the same time ^{14}C-leucine is added. After 10 minutes at 42°, trichloracetic acid is added and the acid-insoluble material is isolated and examined; it is very radioactive. If, instead, the cells are broken open and the cell extract is fractionated by centrifugation, it is found that all ^{14}C is in a fraction containing ribosomes; there is no free ^{14}C-labeled protein. What process is probably temperature-sensitive?

10-20. Suppose you have a mutant of *E. coli* which grows normally at 30° C but slowly at 42° C. There is nothing obviously defective about the rates of DNA and RNA formation, or protein synthesis shortly after the temperature is increased from 30° to 42°. In studying the induction of the synthesis of the enzyme β-galactosidase with this mutant it is observed that if the inducer is removed, at 30° C the rate of β-galactosidase synthesis decreases with a half-life of 5 minutes. At 42°, however, the rate of synthesis does not decline but the amount synthesized per generation *per cell* decreases with a half-life of one generation. What process is probably inhibited at 42°?

10-21. Excluding the enzymes needed to synthesize the amino acids, make a rough count of the number of known enzymes required to synthesize a single protein molecule. What fraction of *E. coli* enzymes is involved in protein synthesis?

●10-22. Translation has evolved in a particular polarity with respect to the mRNA molecule. What would be the disadvantages of having the reverse polarity?

10-23. (a) Suppose a batch of reticulocytes actively making hemoglobin were given radioactive amino acids for a brief period (where a "brief period" is a time short compared to the time required to polymerize the amino acids of a protein chain). Immediately thereafter, the soluble hemoglobin (that which is not attached to ribosomes) is examined for distribution of radioactivity in the chain. Which one of the following will be observed? (1) Label is found in amino acids at the -NH$_2$ end only; (2) Label is found in amino acids at all positions; (3) Label is found at -COOH end only.

(b) Suppose the reticulocytes were permitted to continue making hemoglobin with nonradioactive amino acids for a long time following the radioactive pulse described in (a). If the soluble hemoglobin were examined, which one of the following would be found? (1) Label at the -NH$_2$ end only; (2) Label in amino acids at all positions; (3) Label at the -COOH end only.

○**10-24.** What are the requirements for the union of one 30s and one 50s ribosomal subunit *in vivo* and *in vitro?*

●**10-25.** Suppose you are studying protein synthesis *in vitro* with the following synthetic polymer: AUGUGUUAG. With your standard preparation of factors and ribosomes you find the expected production of free fMet Lys. If you make the same standard factor preparation after infection by phage ϕX19, you find no free dipeptide, but do find fMet Leu bound to the ribosome. Propose an explanation and an additional experiment to verify your proposal.

11

Suppressors

Introduction

In the study of reversion of mutations in bacteria and in yeast, it was found that reversion is frequently a result of a second mutation, which is not even in the original mutant gene. These mutations were said to be the result of *extragenic suppressors.* It was later found that extragenic suppressors are active against many mutations, and in yeast they were termed supersuppressors. In the study of mutants of several *E. coli* phages it was also observed that a wild-type phenotype resulted when some mutants were propagated on certain bacteria; these mutations were called *conditional mutations* or conditional lethal mutations if unable to form plaques on certain strains. Later it was found that the bacterial strains which are *permissive* (that is, they allow the conditional lethal phage mutants to grow) are those strains which carry extragenic suppressors. From study of phage mutants, a peculiar terminology arose, which has become standard. One class of conditional lethal mutants of *E. coli* phage T4 comprises *amber mutants,* and a bacterium permissive for amber mutations was said to harbor an *amber suppressor.* Mutants of other classes were called *ochre* and *opal* mutants, and these are suppressed in permissive strains containing ochre and opal suppressors.* Five years later it was shown that these mutations exist when a normal codon is changed to one of the three different chain-termination codons and that suppressors are altered tRNA molecules, and an understanding of suppression grew rapidly.

NONSENSE SUPPRESSION

One normally thinks of mutations as occurring only in proteins. However, since tRNA is transcribed from DNA, a mutant tRNA molecule, one having a changed base, can certainly exist. If this base-change were to occur in the anticodon region, a tRNA molecule could be produced that has an anticodon complementary to a chain-termination codon (for which normally no tRNA exists). Such a mutant tRNA could then insert an amino acid (the identity of which would depend upon which tRNA was mutated) wherever there is a chain-termination codon complementary to the mutant anticodon. Suppression, that is, the appearance of the wild-type phenotype, will occur if the particular amino acid inserted by the mutant tRNA molecule can substitute for the amino acid present in the original protein. A tRNA molecule that can suppress a chain-termination mutation is called a *nonsense suppressor.* These fall into three classes according to the nonsense codon which is read— UAG, the amber codon; UAA, the ochre codon; and UGA, the opal codon. There are several suppressors for the UAG and UAA codons, since several different tRNA molecules can mutate to form the necessary anticodon. Unfortunately, the simplicity of the explanation of nonsense suppression is lost when the explanation is applied to the UGA suppressor; only one is known, a mutant tryptophanyl-tRNA molecule which can read both the normal tryptophan codon, UGG, and the opal codon, UGA, and which does not have a base change in the anticodon but elsewhere in the molecule. The nature of this suppressor is not yet understood.

It seems that a cell should not survive if a nonsense suppressor is present, since (1) it would lack a necessary tRNA and (2) it would not terminate its proteins normally. However, neither of these points can be correct since, in fact, cells harboring suppressors do survive. The

---

* Amber mutants were named after one of the people in the group at the California Institute of Technology which isolated them, Harris Bernstein, since *bernstein* translates from German to the English word *amber*. Because amber is yellow, the term *ochre* was selected for the second class, ochre being a yellow clay. The third class was at first called *topaz* (also yellow) but this was replaced by *opal*, presumably by someone who had seen only the yellow Mexican fire opal and not the more common pink and blue varieties. There is no significance to the yellow color (amber mutants are not yellow), the whole naming process beginning as a joke.

situation is best understood with the amber suppressors. For example, there are two different genes coding for tyrosyl-tRNA. One of these is responsible for synthesis of most of the tyrosyl-tRNA and the other for a minor component. It is always the minor gene which is mutated to form the tyrosyl amber suppressor, so there is ample functional tyrosyl-tRNA. The ability of amber suppressors to bind to a UAG codon is very great—50 percent of the UAG codons are replaced by an amino acid. However, the dilemma posed by point (2) is avoided since, in the termination of normal proteins, the UAG codon probably rarely occurs as a single termination codon but is usually adjacent to another, different, termination codon. Thus the second codon still causes normal chain termination and cells harboring UAG suppressors grow normally.

The situation is less clear with the UAA suppressors. These suppressors are inefficient, inserting amino acids at only a few percent of the UAA codons. Furthermore, cells harboring UAA suppressors grow poorly. Presumably many genes terminate with single UAA codons so that the ability to survive a mutation to UAA is costly to the cell in that many proteins are not terminated. This is speculation, however, and more work must be done before UAA suppression is understood.

MISSENSE SUPPRESSION

Missense suppression refers to the replacement of one amino acid by another. A mutant tRNA can also provide missense suppression if the anticodon is altered to read a different codon, the same as with nonsense suppression. There are several other types of mutations which can lead to missense suppression. Normally, a tRNA molecule is charged with the correct amino acid by a particular aminoacyl synthetase, which reacts specifically with one amino acid and carries it to a tRNA having the recognition loop corresponding to that synthetase. A base change can occur in a recognition loop so that some specificity of interaction with the amino acyl synthetase is lost. For example, a mutation in glycyl-tRNA might allow the leucyl synthetase to charge glycyl-tRNA with leucine at low frequency; since the glycyl-tRNA would still have the glycine anticodon, leucine would sometimes replace glycine in protein. Alternatively, a mutation could occur in the gene coding for the glycyl synthetase so that it would sometimes react with leucine, or in the gene for leucyl synthetase so that it could sometimes bind to the recognition loop of glycyl-tRNA.

Missense suppression is usually very inefficient; otherwise a very large number of normal proteins would be altered.

REFERENCES

Garen, A. 1968. "Sense and nonsense in the genetic code." *Science,* **160:** 149–159.

Smith, J. D. 1972. "Genetics of transfer RNA." *Ann. Review of Genetics,* **6:** 235–256. For advanced readers.

Stent, G., and R. Calendar. 1978. *Molecular Genetics.* San Francisco. W. H. Freeman and Company, Chapters 15 and 16.

Stryer, L. 1975. *Biochemistry.* San Francisco. W. H. Freeman and Company. Chapter 26.

Watson, J. D. 1976. *Molecular Biology of the Gene.* Third Edition. Menlo Park, Calif. W. A. Benjamin. Chapter 13.

Problems

○11-1. When revertants are analyzed, it is frequently found that reversion has not occurred by a restoration of the wild-type base sequence but is instead a mutation elsewhere in the same gene. Explain how such reversion could restore activity to a mutant enzyme.

○11-2. What is the difference between a missense and a nonsense suppressor? Which type of suppressor might occur by way of a mutant aminoacyl synthetase?

11-3. Give the experimental evidence for nonsense suppressors being altered tRNA molecules.

11-4. Since nonsense suppressors are mutant tRNA molecules, how does the cell survive loss of a needed tRNA by this mutation?

11-5. An amber suppressor supplies an amino acid at the amber termination triplet, UAG. It would seem as if, in a cell containing an amber suppressor, many necessary proteins might not be terminated, so that possession of this suppressor would be lethal. Why is this not the case?

11-6. Why are ochre (UAA) suppressors usually weak suppressors?

11-7. How might it be possible for an extragenic suppressor to suppress a frameshift mutation?

○11-8. A missense suppressor substitutes a particular amino acid for a "wrong" amino acid in a mutant protein. Describe two ways that missense suppression could occur. Why are missense suppressors usually weak? If a missense suppressor substitutes amino acid Y for amino acid X, would you expect it to suppress all mutations in which X is the "wrong" amino acid in the mutant? Explain.

11-9. Consider two genes *a* and *b*. In growth medium Q, a mutation in either *a* or *b* is lethal (that is, the genotypes a^-b^+ and a^+b^- are lethal combinations. A genetic cross is made between one bacterium with genotype a^-b^+ and a second with genotype a^+b^-. It is found that there are two types of recombinants that can grow in Q, the a^+b^+ wild-type and (surprisingly) the a^-b^- double mutant—that is, a^- suppresses the b^- mutation and b^- suppresses the a^- mutation. Propose a molecular explanation for this.

11-10. Suppressors were originally recognized by their ability to suppress a large number of mutants of many types. Frameshift suppressors differ in that usually only a small number of frameshift mutations are suppressed. A necessary but not sufficient property of the mutation is that it is a *single* base addition. It is known that these suppressors are tRNA molecules which have been altered. Propose a way in which a tRNA could be altered to become a frameshift suppressor. If you wanted to mutate *E. coli* to produce a frameshift suppressor, which mutagen would you use? Do you think a frameshift mutation consisting of a single base deletion could be suppressed? Explain.

11-11. Consider a piece of DNA in the nonessential part of the T4 protein specified by the *rII* gene. (See Figure 11-1. Note that only the strand from which the messenger RNA is copied is shown. The other is irrelevant for this problem.) Suppose that, by acridine orange treatment, a new sequence is created so that two G's are inserted at the point indicated. This will inactivate the protein because reading will be out of phase. However, if the G at the second indicated point is then mutated to become a C, an active, *rII* protein (but not necessarily the same one) is made. Explain.

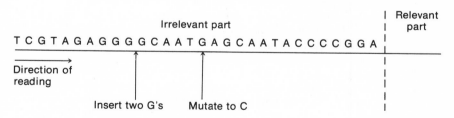

FIGURE 11-1

● 11-12. Devise a method for the isolation of temperature-sensitive amber suppressors.

● 11-13. One of the first isolated *E. coli* mutants, in which DNA synthesis is sensitive to temperature change, carries a mutation called *dnaB*266. This mutation was found to be 30 percent contransducible with a *mal* gene (maltose fermentation), when P1-phage transduction was carried out as follows. The donor bacterium had the genotype *mal*⁺ *supE dnaB*266 and the recipient bacterium had the genotype *mal*⁻*supE dnaB*⁺; P1 phage was grown on the donor strain and used to transduce the recipient strain. After P1 infection of the recipient strain, the infected cells were spread at 30° C on plates containing maltose as the sole carbon source, and the *mal⁺* colonies which appeared were picked and tested for ability to grow at 42° C. Thirty percent of the colonies were unable to grow at the restrictive temperature. Consider the following results of an attempt to transduce the *dnaB*266 mutation into a variety of recipient strains: (1) When the *mal⁻* recipient carried *supD*, then *dnaB*266 was 30 percent contransducible with *mal⁺*. (2) When the recipient was nonsuppressing, none of the transductants were temperature-sensitive. (3) When the recipient carried *supF*, none of the *mal⁺* transductants were temperature-sensitive. However, 30 percent of the *mal⁺supF* transductants must carry the *dnaB*266 mutation, because these transductant strains can be used as donors in a second transduction, and can transfer a temperature-sensitive mutation into an *E. coli* having a *supE mal⁻* genotype. Explain these findings in terms of protein structure and the action of suppressors.

11-14. Explain the following two phenomena.
(a) A suppressed missense or nonsense mutation often has a mutant phenotype at 42° C but not at 30° C even when it is known independently that the suppressor itself functions equally well at both temperatures.
(b) A mutant protein has amino acid B at a site normally occupied by A. When the mutant gene responsible for synthesis of the mutant is transduced into a bacterium containing a missense suppressor, some activity of the protein is restored; however, it is known that the particular suppressor substitutes amino acid D for C and never substitutes A for B.

●11-15. It is sometimes found that a particular base-pair deletion, that is, frameshift mutations designated as −, is suppressed if followed by any of several different base-pair insertions, designated by +, but some of the + mutations, if followed by a different − mutation, are not suppressed, even though the mutant and suppressive base-pair changes occur in the same region of the cistron. What could be the reason for this finding and how would you test your hypothesis?

○11-16. A frameshift mutation consisting of addition of a single base could be corrected by which of the following?
(a) A mutant aminoacyl synthetase.
(b) A tRNA molecule containing an anticodon capable of recognizing a chain-termination mutation.
(c) A tRNA molecule containing an anticodon four bases long.
(d) An enzymatic system which could chemically modify a particular amino acid.

●11-17. Generally, suppressor mutations for the UAA type of nonsense mutation also suppress the UAG type of nonsense mutation. Why do you think that this occurs? Suppose you decide to look into this problem more thoroughly by isolating a larger set of UAA suppressor mutations. After extensive mutagenesis, you isolate some rare "restricted"-range UAA-suppressor mutations which suppress UAA and not UAG. You now also find suppression of certain missense mutations. Propose an explanation in terms of tRNA structure. What amino acid do you predict will be replaced by the latter class of suppressor tRNA molecules?

12

Enzyme Structure and Regulation

Introduction

In Chapter 8 we discussed how the activity of an enzyme is regulated at the level of transcription of the DNA coding for the enzyme—that is, enzyme synthesis is turned on or off according to whether the enzyme is needed by the cell. Some enzymes are synthesized continuously and in these, regulation consists of inactivation or activation of the enzyme. Studies of regulation of this type invariably require a detailed analysis of the chemical and physical properties of enzymes and how these properties are affected by the binding of the substrate or other ligands.

This subject is too broad to be treated comprehensively within the scope of this book, and the problems that have been included are for students familiar with the theory of ligand binding, enzyme kinetics, and the energetics of protein folding. A few words seem appropriate, however, concerning two enzymatic phenomena which are often considered by molecular biologists—*feedback inhibition,* and *allostery.*

FEEDBACK INHIBITION

Consider the following biosynthetic pathway:

$$A \xrightarrow{1} B \xrightarrow{2} C \xrightarrow{3} D,$$

catalyzed by enzymes 1, 2, and 3. If D is provided ready-made to a cell, there is no need for the cell to maintain this pathway; in fact, it might be disadvantageous for the cell to do so if A is needed in another pathway. Therefore, it is common for enzyme 1 to have two binding sites, one for A and the other for D. When D is bound to enzyme 1, the enzyme is unable to catalyze the conversion of A to B, usually because the (D:enzyme 1) complex is unable to bind A. This is the simplest kind of feedback inhibition.

In the preceding example inhibition of enzyme 2 by D would also prevent synthesis of C and D. This would be wasteful unless there is some reason to maintain the reaction catalyzed by enzyme 1. This is the case if the pathway is branched and B is needed in two different syntheses. As an example of such a situation, consider the branched biosynthetic pathway

$$A \xrightarrow{1} B \begin{array}{c} \nearrow^{4} E \xrightarrow{5} F \\ \searrow_{2} \\ C \xrightarrow{3} D \end{array}$$

In this pathway the most economic kind of inhibition again usually prevails: D inhibits enzyme 2 and F inhibits 4; however, inhibition of enzyme 1 by either D or F would clearly be deleterious. On the other hand, conversion of A to B in this branched pathway is wasteful if D and F are *both* present. If both D and F are present, inhibition of enzyme 1 can be achieved in several ways: (1) inhibition of enzyme 1 by B, which accumulates when enzymes 2 and 4 are blocked; (2) enzyme 1 has binding sites for both D and F and when both binding sites are filled, the enzyme is inactivated; (3) D and F both inhibit enzyme 1 slightly, for example, by D-fold and F-fold degrees respectively, and together they inhibit the enzyme $(D \times F)$-fold degrees; (4) in some pathways, A is converted to B by two different enzymes *(isozymes)*, one inhibited by D and one by F.

There are many ways in which the enzyme inhibition might occur. A common mechanism is allosteric change.

ALLOSTERY

Allostery refers to a phenomenon in which the binding of one molecule to a particular active site on a protein alters the affinity of another active site for a second substance.* Presumably this occurs because binding of the first molecule induces a conformational change in the protein.

Allosteric changes can result in both activation and inhibition of an enzyme. The examples given so far have all dealt with inhibition. We now present two commonly occurring mechanisms for activation. (1) Some enzymes have a binding site for a small activator molecule and a potential (inactive) site for the substrate. The activator binds to the enzyme and induces a conformational change which converts the potential binding site to an active binding site. (2) Some enzymes have two weak binding sites for the substrate. When one molecule of substrate is bound, a conformational change occurs and the second weak site is converted to a strong binding site. An important variant of the second kind of activation is for the enzyme to consist of two identical subunits, each with one binding site; when a substrate molecule is bound to one subunit, this subunit changes its shape and thereby induces the second subunit to change so as to be in a strong binding form. Mechanism (2) is important in cell economy, since the enzymatic activity described by this mechanism responds directly to the substrate, the fraction of active enzymes increasing with substrate concentration.

The detailed molecular mechanism for allosteric activation has not been worked out. Two models dealing with allosteric changes in multi-subunit proteins have been proposed—the *symmetry* model, or *concerted* model of Jacques Monod, Jeffries Wyman, and Jean-Pierre Changeux, and the *sequential* model of Daniel Koshland. Both of these models are based upon the ideas that proteins are flexible in conformation and that a protein subunit can exist in only two states. The models differ in several points through. First, the symmetry model assumes that the active and inactive states are in equilibrium and the sequential model assumes that the transition from inactive state to active state is *induced* by substrate binding. Second, the symmetry model states that if one subunit changes its form, the other subunit necessarily changes to that form also because, unless symmetry is preserved, the two subunits cannot interact. The sequential model allows interaction between subunits in different states. Third, the symmetry model assumes that binding of the first molecule neces-

---

Allostery is derived from the Greek words *allos* (another) and *stereos* (solid).

sarily enhances binding of a second molecule of the same type; the sequential model allows either an increase or decrease of affinity for the second.

It is difficult to make a strong statement about the relative validity of the two models. Some enzymes are best described by the first, others by the second. One group of allosteric proteins, in which there may be more than two states, is described by neither.

REFERENCES

Atkinson, D. E. 1969. "Regulation of enzyme function." *Ann. Rev. Microbiol.*, **23**: 47–68. Contains a discussion of allostery.

Bernhard, S. A. 1968. *The Structure and Function of Enzymes.* Menlo Park, Calif. W. A. Benjamin. An introductory text.

Changeux, J.-P. 1965. "The control of biochemical reactions." *Scientific American*, **April**: 36–45.

Koshland, D. E. 1973. "Protein shape and biological control." *Scientific American*, **October**: 52–64. A simple description of conformational changes.

Koshland, D. E., G. Nemethy, and D. Filmer. 1966. "Comparison of experimental binding data and theoretical models in proteins containing subunits." *Biochemistry*, **5**: 365–385. An advanced paper in which the sequential model was first stated.

Lehninger, A. L. 1974. *Biochemistry.* New York. Worth Publishers, Inc. Second edition. Chapter 9.

Monod, J., J.-P. Changeux, and F. Jacob. 1963. "Allosteric proteins and cellular control systems." *J. Mol. Biol.*, **6**: 306–329. The first paper in which the concept of allostery was introduced.

Monod, J., J. Wyman, and J.-P. Changeux. 1965. "On the nature of allosteric transitions." *J. Mol. Biol.*, **12**: 88–118. Presentation of the symmetry model.

Stent, G., and R. Calendar. 1978. *Molecular Genetics.* San Francisco. W. H. Freeman and Company. Chapter 4. For the beginner.

Stryer, L. 1975. *Biochemistry.* W. H. Freeman and Company. Chapter 6. An excellent exposition for the beginner with very clear diagrams.

Wold, F. 1971. *Macromolecules: Structure and Function.* Englewood Cliffs, N. J. Prentice-Hall. Chapters 2 and 8. Somewhat advanced.

Problems

12-1. Many enzymes are aggregates of identical subunits. Explain the biological advantages of this phenomenon.

12-2. Why do enzymes vary so much in size? Would you expect the enzymes involved in glucose metabolism to be relatively large or relatively small? What would you expect to be the size range of the various DNA polymerases?

12-3. The regulation of the synthesis of an enzyme can be accomplished at the level of transcription or translation. Suppose you are studying the kinetics of appearance of an inducible enzyme after the addition of an inducer. You can assay the enzyme both by its enzymatic activity and by an immunological test which shows that the protein is present. After induction you observe that enzymatic activity does not appear until several minutes after the appearance of immunological reactivity. Give several possible explanations for this.

12-4. No amino acid side chain is a good electron acceptor, yet many enzymes catalyze reactions in which the enzyme serves as an electron acceptor. How is this accomplished?

12-5. The activity of many enzymes is strongly dependent on pH and shows maximum activity at a particular pH. Furthermore, the range of pH values at which there is significant activity is often only two to three pH units: often there is no detectable conformational change in this pH range so that it is unlikely that a gross conformational change accounts for this behavior.
(a) Explain this dependence on pH.
(b) What might you conclude about the active site of an enzyme if the activity increases from pH 5 to pH 7 and then remains constant (that is, it has no maximum) until pH 11, at which value the activity abruptly drops to zero?

○**12-6.** Consider the branched pathway indicated below, which is regulated by feedback inhibition.

$$A \xrightarrow{1} B \xrightarrow{2} C \xrightarrow{3} D \xrightarrow{4} E \xrightarrow{5} F \xrightarrow{6} G$$

$$ {}^{8}\downarrow {}^{7}\downarrow$$

$$ I H$$

$$ {}^{9}\downarrow$$

$$ J$$

(a) Indicate the enzymes which are subject to feedback inhibition and, for each, identify the inhibitor.

(b) Which steps are likely to be catalyzed by a set of isozymes? How many isozymes are probably in these steps?

(c) Assume that isozymes are not involved in this reaction scheme. Propose an alternative mechanism for inhibiting the steps discussed in part (b).

12-7. Regulation of enzyme activity usually involves modifications of the protein molecule (as, for examples, in competitive inhibition, allosteric inhibition, and so on). However, it is also possible that the substrate is modified to make it more or less available to an enzyme. Give examples of substrate modifications which affect

(a) The ability of DNA to be transcribed.

(b) The sensitivity of DNA to nucleases.

(c) The ability of DNA to engage in various types of genetic recombination.

12-8. The addition of a low concentration of a competitive inhibitor to certain enzymes that have an active and an inactive form sometimes results in activation. Explain why.

●12-9. In a paper by Monod, J., J. Wyman, and J. P. Changeux, *J. Mol. Biol.* (1965), **12**, 88–118, the following equation is given to describe interactions between identical ligands (homotropic effects) in binding by an allosteric protein:

$$\bar{y}_F = \frac{L_0 c \alpha (1 + c\alpha)^{n-1} + \alpha(1 + \alpha)^{n-1}}{L_0(1 + c\alpha)^n + (1 + \alpha)^n},$$

in which $\bar{y}_F$ is the fraction of sites actually found by the ligand, $\alpha = F/K_R$, $c = K_R/K_T$, n is the number of potential binding sites for the ligand, F is the ligand concentration, L_0 is the equilibrium constant for the transition of the protein from the R (relaxed) state to the T (tense) state, and K_R and K_T are the dissociation constants for the binding of the ligand to a site in the R and T states, respectively.

(a) Reduce this equation to its simplest form for the following conditions: (1), $c = 0$; (2), $c = 1$; (3), $L_0 = 0$; (4), $L_0 = \infty$, $c > 0$; (5), $n = 4$; (6), $n = 1$. Then give a qualitative comparison of the cooperativity possible in ligand binding under each of the conditions given above.

(b) For a tetramer, let $L_0 = 10^4$ and $c = 0.01$. Evaluate L_1, L_2, L_3, and L_4, that is, the ratio of T to R states when 1, 2, 3, and 4 ligand molecules are bound per protein molecule, in the two states.

12-10. Consider an allosteric protein which exists either as a monomer or as a tetramer. A real example would be the isolated α subunit of hemoglobin versus hemoglobin itself. Both forms of the protein bind the ligand, F. Draw out the saturation curves for these two proteins, using the Monod-Wyman-Changeux equation of problem

12-9, with the parameters evaluated as shown in Table 12-1. In order to compare the amplifying properties of these two proteins, compare the change in concentration of F required to give a change in $\bar{y}$ (the fractional saturation) from 0.1 to 0.9, in the two cases.

TABLE 12-1

Parameter	Monomer	Tetramer
n	1	4
c	0	0
L_o	10	10^4

12-11. A new enzyme catalyzing the first step in the synthesis of vitamin Z and a mutant enzyme have been isolated. The molecular weights (determined by sedimentation analysis of the wild-type and mutant enzymes) are 100,000 and 25,000, respectively. Both enzymes are inhibited by vitamin Z. The enzyme kinetics (that is, reaction velocity, V, versus substrate concentration, $[S]$) differ as shown in Figure 12-1; $+Z$ and $-Z$ refer to the presence and absence of vitamin Z.

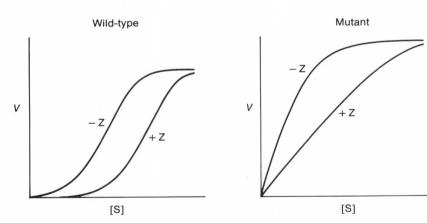

FIGURE 12-1

(a) Make a drawing which explains the kinetics of the mutant and wild-type enzymes *in the absence* of vitamin Z. Your drawing should indicate the role played by the binding of substrate, and conformational changes.

(b) Make another drawing which explains how vitamin Z inhibits both the wild-type and mutant enzymes. Indicate binding sites and conformational changes (if any).

(c) Suppose you introduced, by an F' factor, the gene for the mutant enzyme into a wild-type strain, creating a partial diploid strain. If you isolate the enzyme from these cells, would you expect to see a "poison subunit" effect? Explain.

12-12. Recently a protein that binds cyclic AMP (cAMP) was isolated from *E. coli* and characterized by equilibrium dialysis experiments. This protein has a molecular weight of approximately 6×10^4 as determined by sucrose-gradient zone-centrifugation. Given in Table 12-2 are raw data from an equilibrium dialysis experiment in which protein was present inside the dialysis bag at 1.2 mg/ml. The cAMP concentration is reported as total concentration (bound and unbound) inside the bag. Assume protein does not pass through the bag while cAMP does.

TABLE 12-2

Molarity of cyclic AMP	
Outside bag	*Inside bag*
0.7×10^{-6}	2.7×10^{-6}
2.3×10^{-6}	8.3×10^{-6}
4.5×10^{-6}	1.5×10^{-5}
1.0×10^{-5}	2.6×10^{-5}
3.0×10^{-5}	5.4×10^{-5}
7.0×10^{-5}	9.8×10^{-5}
1.5×10^{-4}	1.8×10^{-4}

(a) Determine K_D, the dissociation constant for the cAMP-protein complex.

(b) Determine n, the number of binding sites per protein molecule, of the assigned molecular weight.

(c) If you get a nonintegral number of sites, what explanations can you offer to account for this result?

12-13. Diisopropylfluorophosphate (DFP) reacts covalently with the serine hydroxyl group found at the active site of many proteases and esterases, and enzyme activity is irreversibly lost. The inactivation reaction follows pseudo-first-order kinetics when DFP is present in excess. Analyze the data in Table 12-3 and determine the pseudo-first-order rate constant and predict the time required to obtain 99.9 percent inactivation of the enzyme.

TABLE 12-3

Time (minutes)	Percent of activity remaining
0	100
2	73.5
4	54.0
8	29.5
16	8.5

●12-14. There are temperature-sensitive mutants of *E. coli* capable of growth at 30° C but not at 42° C, in contrast to the wild-type, which grows at 42° C as well. Such mutants are thought to possess some enzyme that serves a critical function and that, even though synthesized at both temperatures, becomes inactive at high temperature. Discuss in a few sentences the characteristics of a protein-folding reaction which would account for the growth behavior of these mutants—that is, why might the temperature dependence of this enzymatic activity be so great?

●12-15. *E. coli* tryptophan synthetase in its catalytically active form contains two nonidentical polypeptide chains, α and β. A single bacterium of volume 1 μm^3 is estimated to contain about 100 copies of each chain. In order for 90 percent of the chains to exist in the $\alpha\beta$ complex, what value of K_D must there be for the complex?

●12-16. Most metabolic intermediates and macromolecules are liable to spontaneous hydrolytic destruction. The stability of such a compound is described by its free energy of activation, ΔG. Consider a hypothetical set of molecules with the ΔG values for hydrolysis of 7, 14, 28, and 42 kcal/mole and *calculate* the half-life of each in aqueous solution at 37° C.

TABLE 12-4

Strain	K_M	V_{max}
Wild-type	$10^{-3}\ M$	100
Mutant 1	$10^{-4}\ M$	10
Mutant 2	$10^{-2}\ M$	250
Mutant 3	$10^{-2}\ M$	2

●12-17. β-Galactosidase from *E. coli* has been studied with respect to the effect of mutation on its catalytic activity. The K_M and V_{max} characteristics of three of these mutant enzymes are shown in Table 12-4. This enzyme catalyzes the hydrolysis of the disaccharide lactose by a mechanism fully analogous to the lysozyme mechanism for the hydrolysis of NAG (*N*-acetyl glucosamine) polymers. From your understanding of lysozyme, propose what features of the hydrolytic mechanism of β-galactosidase must have been modified in the mutants in order to cause the changes in K_M and V_{max}. Assume that K_M values are equivalent to K_D values.

13

Repair and Repair Replication

Introduction

DNA molecules are subject to continual attack by radiation and chemicals. In order to survive, systems have evolved which can repair DNA damage. These systems recognize chemically altered bases, mismatched base pairs, and strand breaks, and they restore the *status quo* either by direct chemical repair, excising the damaged parts and re-synthesizing "correct" DNA, or by recombining functional parts of the DNA and discarding the damaged regions. At the present time many repair mechanisms are known, and it is likely that more remain to be discovered.

SURVIVAL CURVES AND THE DOSE-REDUCTION FACTOR

The extent of damage to cells is usually expressed with a survival curve, where survival can mean retention of the ability to grow (that is, to form a colony or a plaque) or the ability to carry out a specific process such as an enzyme reaction. A survival curve is usually plotted in terms of the logarithm of the surviving fraction as a function of dose (degree of exposure to a damaging agent), since in certain cases the curve will be a straight line. If damaging events (*hits*) are randomly distributed over a population, and if only a single hit causes inactivation, the equation $N = N_0 e^{-kD}$ describes the inactivation, where N is the number of active units (survivors), N_0 is the number of initially active units, k is a constant depending on the frequency with which potential damage becomes real damage (called "efficiency"), and D is the dose. Hence $\ln(N/N_0) = -kD$ and a plot of $\ln(N/N_0)$ versus D yields a straight line. A simple example of this type of kinetics is the loss of plaque-forming ability in an X-irradiated phage population. The X-rays cause, among other things, double-strand breaks, and each double-strand break is a lethal hit.

Let us now consider the situation in which two hits are necessary. A simple example would be an agent that produces single-strand breaks in DNA. These are not lethal until two on opposite strands match to form a double-strand break. Thus a survival curve in this case will have a shoulder—that is, there will be no initial inactivation until many hits accumulate in the population. In the very simple case of a two-hit curve, an extrapolation of the so-called linear portion of the curve (which in fact is never linear but asymptotic) to the vertical axis yields an intercept of 2. In the general case of an n-hit curve, this intercept is n. (See Figure 13-1.)

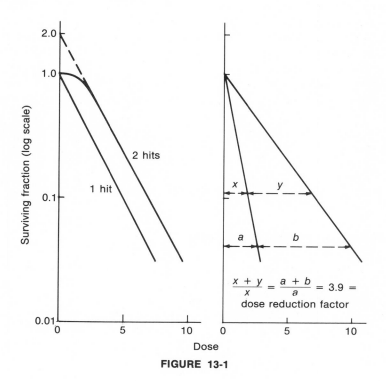

FIGURE 13-1

Now let us assume that a repair system exists so that some of the damage is eliminated. With respect to dose, the inactivation will proceed at a lower rate than if there were no repair. Consider a case in which there are two kinds of damage, one occurring at twice the frequency of the other and being repaired at 100 percent efficiency. Therefore, three times the dose which is required when there is no repair will be needed to produce the same surviving fraction. It is then said that the repair (or, more generally, a process preventing permanent damage) introduces a *dose-reduction factor* of three. Note that the same dose-reduction factor would result if there were only a single type of damage which is repaired at an efficiency of $\frac{2}{3}$.

At this point it is necessary to note a common error encountered in discussing the efficiency of different repair processes. Suppose an irradiated population is given a dose D which, in the absence and presence of repair, gives 50 percent and 75 percent survival, respectively. The survival curve is a single-hit graph. In a sense, one can (incorrectly) think of this in terms of a two-fold change in survival—that is, 50 percent are killed when there is no repair and 25 percent when there is repair. However, the repair efficiency is not 2; if another dose were chosen so that survival without repair was 12.5 percent, repair would not yield 25 percent but 42.18 percent. The correct way to state the efficiency of repair is to designate the dose-reduction factor (the ratio of the doses with and without repair required to yield the same survival). This is most easily done by plotting the survival curves, drawing a horizontal line from the vertical axis of each curve and measuring the ratio of the distances from the vertical axis to each intercept. Clearly this ratio will be constant at all survival levels, for both single-hit and n-hit curves.

Frequently, repair cannot be expressed simply in terms of dose-reduction factors, since the *shape* of the survival curve can change. This could occur, for instance, if the repair process itself were susceptible to damage or if it were induced by the presence of damage and its effectiveness was related to the amount of dosage.

REPAIR OF PYRIMIDINE DIMERS

Ultraviolet (UV) irradiation produces chemical linkage between adjacent thymines or adjacent cytosines on the same strand and forms a compound called a *pyrimidine dimer.* (Linkage between thymines on different strands also occurs but the frequency is lower by many orders of magnitude.) At present, repair of pyrimidine dimers is the best-understood repair process. In the following, some of the mechanisms for repair of these dimers are briefly described.

1. Photoreactivation. Pyrimidine dimers are cleaved to re-form the monomeric structure by an enzyme which is activated by the absorption of blue light. The chemical mechanism by which this occurs is not known. Photoreactivation introduces a dose-reduction factor of approximately two in the survival of bacteria and yeasts; this indicates either that lesions other than pyrimidine dimers are produced by the UV light or that all dimers cannot be cleaved by photoreactivation (perhaps certain adjacent base sequences inhibit enzyme binding).

2. Dark repair or excision repair. If a population of irradiated bacteria is prevented from growing for a period of time after irradiation, the likelihood of survival is much greater. This is called *liquid-holding recovery.** Similarly, if an irradiated phage is allowed to inject its DNA into a bacterium but not to replicate for a short while, again the likelihood of survival of the phage is greater. In contrast to photoreactivation (the first repair system discovered), the type of repair not requiring blue light yet effective as soon as damage has been done was called *dark repair.* It occurs even if growth is not experimentally delayed, beginning as soon as pyrimidine dimers appear. Dark repair is accomplished in four steps: first, *incision,* by which a specific endonuclease called a UV-endonuclease makes a single-strand break near the dimer, usually to the 5' side; second, *repair replication,* by which a DNA polymerase initiates growth from the normal segment of DNA and displaces a segment containing the pyrimidine dimer; third, *excision,* by which a nuclease removes the segment carrying the dimer; fourth, *sealing,* by which the newly synthesized segment is joined to the main body of the DNA by DNA ligase. In some cases, excision can precede repair replication. The general mechanism for excision repair is called the *"cut, patch, cut, and seal"* or *"cut-and-patch" model.*

Excision repair is active not only against pyrimidine dimers but generally against distortions in the helix produced by alteration of the bases by ionizing radiation or by alkylation of purines. Specific nucleases (not the UV-endonuclease) are required for incision of each of these lesions.

3. Recombination repair. If excision repair is prevented by mutation or if some dimers fail to be either photoreactivated or excised, lesions can still be eliminated by *recombination repair.* This process is poorly understood but can be outlined as follows. If DNA containing

---

* This phenomenon was originally detected by holding irradiated cells in a nonnutrient liquid buffer during which time the cells seemed to recover from the deleterious effects of the irradiation.

pyrimidine dimers is replicated, difficulties are encountered when the replication fork meets the dimer. In a way that is not fully clear, daughter DNA molecules containing gaps result. These are deleterious to the organism since, while they are duplicated, each gapped template-strand will produce fragments of double-stranded DNA and each continuous strand containing dimers will make more gapped DNA. In recombination repair, two gapped daughter DNA molecules are broken and rejoined (perhaps stimulated to do this by the gaps) and DNA molecules free from damage often result. Recombination repair is the preferred mechanism for repairing DNA containing cross-links between the two single strands.

EFFICIENCY OF REPAIR

Together, dark repair and recombination repair introduce a dose-reduction factor of about 10^3—clearly these are two highly efficient systems. However, with very large amounts of damage, cells and phages are still killed; presumably this failure to survive represents either saturation or inactivation of the repair systems, or the production of nonrepairable damage.

REPAIR AND MUTATION

Damage to DNA often results in mutation. Sometimes this is a direct consequence of chemical alteration of a base so that the base has different hydrogen-bonding properties—an example is the deamination caused by exposure to nitrous acid. However, it often results from errors in repair replication; this may be the principal mechanism of radiation-induced mutation. Presumably the fidelity of repair replication is not as great as that of normal DNA replication; this so-called *error-prone replication* is not yet understood. A particular type of error-prone replication is called **SOS repair.** The name refers to the fact that its primary aim is survival without regard for restoring the original base sequence.

REFERENCES

Giese, A. C. 1963–1971. *Photophysiology.* Vol. 1–6. New York. Academic Press. There are numerous articles on repair in these volumes.
Hanawalt, P. C. 1972. "Repair of genetic material in living cells." *Endeavor,* **31:** 83–87. A well-illustrated review showing the basic elements of repair.

Howard-Flanders, P. 1968. "DNA repair." *Ann. Rev. Biochem.*, **37**: 175–200. An excellent review describing many of the important experiments.
Kornberg, A. 1974. *DNA Synthesis*. San Francisco. W. H. Freeman and Company. Chapter 9. An excellent description of the enzymology of repair.

Problems

○13-1. Which of the following is a true statement about DNA repair?
(a) Thymine dimers are usually formed between adjacent thymines in the same strand.
(b) Thymine dimers are usually formed between thymines in different strands.
(c) The photoreactivation enzyme cleaves all thymine dimers in a UV-irradiated cell.

13-2. What is the purpose of light in the functioning of the photoreactivating enzyme? What common biochemical reaction is replaced by the light?

○13-3. Suppose you had a repair system capable of removing half of the damage produced by some agent. Which of the following would be true of (1) single-hit and (2) multi-hit survival curves?
(a) A dose yielding 40 percent survival in the absence of repair would yield 80 percent survival if repair occurs.
(b) The dose required for n percent survival in the absence of repair is doubled when repair occurs.

●13-4. Consider a bacterium repair system called X, which removes thymine dimers. You have in your bacterial collection the wild-type (X^+) and an X^- mutant. Phage ϕ, when UV-irradiated and then plated, gives a larger number of plaques on X^+ than on X^-. It has been proposed on the basis of survival curves that the X enzyme is inducible. However, X does not work *in vitro* and the enzyme has not been isolated. To investigate this, UV-irradiated ϕ phages are adsorbed to both X^+ and X^- bacteria in the presence of the antibiotic chloramphenicol (which inhibits protein synthesis). No thymine dimers are removed with X^- and 50 percent are removed with the X^+. In the absence of chloramphenicol, the same results were obtained.
(a) Is X an inducible system?
(b) Suppose 5 percent of the thymine dimers are removed in the presence of chloramphenicol and 50 percent in its absence; how would your conclusion be changed?

●13-5.   Consider a bacterium whose DNA is labeled with ^{3}H-thymidine and then UV-irradiated. If the cells are exposed to visible light for a long time, a maximum of 30 percent of the dimers are cleaved (by photoreactivation, abbreviated as PR). If they are incubated in the dark for a long time in a dilute buffer (in order to prevent cell division), 40 percent of the dimers are cleaved. This is called dark repair or DR. In order to determine whether the set of dimers cleaved by PR overlaps the set cleaved by DR, experiments are done in which the two repair systems are allowed to act sequentially and the percent of thymine dimers *remaining* is measured. Listed below are the results of four possible but different experiments (a–d). In each, what conclusion could you draw about the properties of the two repair systems?

Sequence	*Percent remaining dimers*
(a) PR → DR	30
(b) PR → DR	45
(c) DR → PR	45
(d) PR → DR or DR → PR	40 (by both sequences)

13-6.   What is the phenotype of each of the following *E. coli* mutants?

uvrA$^-$, uvrB$^-$, uvrC$^-$, lon$^-$, lexA$^-$, hcr$^-$.

●13-7.   Some years ago, Evelyn Witkin isolated from a heavily UV-irradiated population of *E. coli* B a mutant called B/r, which is resistant to UV light when compared to the parent strain B. Whereas the existence of this mutant among the surviving cells is not surprising, it is more difficult to understand why Ruth Hill found among the survivors of a similar irradiated culture a highly radiosensitive strain known as B$_s$. Propose a mechanism to explain this latter discovery. Would you expect the original colonies of both B/r and B$_s$ to have been pure or to have contained parental wild-type cells?

13-8.   *E. coli* polymerase I mutants are more sensitive to UV irradiation than the parent strain. However, the sensitivity is not as great as that of *uvrA$^-$* strains. In view of the proposed gap-filling role of polymerase I in repair, explain the relative sensitivities of these mutants.

13-9.   An unexcised thymine dimer produces a partial block to DNA replication. It is observed that if there are unexcised thymine dimers in parental strands, the daughter strands contain large gaps which are frequently several thousand nucleotides long.
(a) Would you expect to find gaps in both daughter strands if there

were only a single thymine dimer in the parent molecule? If not, in which strand? Explain.

(b) In an unirradiated cell, if RNA synthesis is blocked by rifampicin or by uracil starvation of a uracil-requiring mutant, DNA molecules in the act of replication can be completely replicated, but a second round of replication cannot begin. Would this also be true for a heavily UV-irradiated cell? Would addition of rifampicin have the same effect as uracil starvation? Explain.

13-10. Why would you expect that a large dose of UV irradiation would kill wild-type cells even if the dose is not large enough to saturate the repair apparatus?

●13-11. In $uvrA^-$ or $uvrB^-$ E. coli strains exposed to UV radiation, the survivors can tolerate the pyrimidine dimers. One hypothesis states that the gaps in the newly synthesized DNA can be filled in by some as yet unknown mechanism; another idea is that the newly synthesized DNA recombines with parental DNA that has unexcised lesions produced by UV irradiation. How would you go about showing whether or not the gap-filling mechanism (which would generate newly synthesized DNA free of pyrimidine dimers) does take place? (For information helpful in working this problem, see *Mol. Gen. Genetics* (1975), **141**: 189.)

●13-12. It is possible to induce a mutation in $polA^-$ strains with smaller doses of UV radiation than are needed to induce a mutation in wild-type E. coli strains. Why? Would you expect the same type of result if you compared the dose of UV irradiation required to induce a mutation in $uvrA^-$ cells with that required for wild-type cells? See Witkin, E., *Genetics* (1975), **79**: 199. What is the best evidence that has been obtained for the inducibility of the SOS system of DNA repair and related activities?

13-13. Suppose you are testing the possibility that an inducible, post-replicative repair system of E. coli has a small molecule effector analogous to cAMP or ppGpp. Assuming that the effector might

TABLE 13-1

Strain	pAp level	
	No UV	*UV*
Wild-type	Low	High
*recA*1	None	None
*lexB*1	None	None
*lexA*1	Low	Low
*lexA*11	High	High

sense a slowdown or shutdown in DNA replication (this is known as the "idling hypothesis"), you look at deoxynucleotide pools and find that after UV irradiation pAp and pGp accumulate in the cells. As a means of slowing down DNA replication, you use a *dnaB-ts* mutant at a nonpermissive temperature and find that pAp, pGp, pCp, and pTp accumulate. In a second study of mutants defective in the repair pathway you obtain the data shown in Table 13-1. All mutations are recessive except for *lexA* 1, which is dominant. The location of *recA* and *lexB* genes are adjacent; *lexA* is far away from these genes.

(a) Provide an explanation of the small-molecule difference between UV inhibition and *dnaB⁻* inhibition of replication; your explanation should preserve the idling hypothesis.

(b) Propose a model for regulation of the "post-replicative repair operon" using the data in the table and the relative positions of the genes.

14

Techniques

Introduction

As is frequently the case, except for a small number of monumental intellectual insights that revolutionize a scientific field, progress follows closely behind technical developments. This has been especially true of molecular biology and biochemistry. Some examples of these developments and their consequences follow.

The invention of the *analytical ultracentrifuge* in the 1920's by The Svedberg provided the first method for detecting inhomogeneity of the molecular weight of macromolecules. This instrument also allowed one to determine the minimum number of types of macromolecules in biological fluids (for example, the number of proteins in blood serum). As new optical systems were designed for the ultracentrifuge, there became available a precise procedure for determining the molecular weights of a macromolecule, *equilibrium centrifugation.* An important variation, *equilibrium centrifugation in a density gradient* (commonly called *density gradient centrifugation*) gives efficient separation of molecules having different densities, and density became a new label in biological experiments. This technique was used in the elegant Meselson-Stahl experiment, which gave important information about the mechanism of DNA replication (see chapter 5). From the analytical ultracentrifuge there evolved the *high-speed prepara-*

tive ultracentrifuge with which macromolecules and organelles, such as ribosomes, chloroplasts, and nucleoli, can be isolated and purified. The preparative centrifuge was used in an analytical modification, *zonal centrifugation in a preformed density gradient,* which allows the technology of radioactive tracers to be joined with sedimentation techniques. The ultracentrifuge made the study of macromolecules feasible and there are now very few biological laboratories that do not possess the instrument or at least have it available.

Another striking example is *chromatography* and its many forms. Chromatography has been known for 2,000 years, having been used by Pliny, but did not make a real impact on biology and chemistry until the development of *partition chromatography* in columns and *adsorption chromatography* on paper in the mid 1940's and early 1950's. Using paper chromatography, Erwin Chargaff determined the base composition of DNA, refuted the tetranucleotide hypothesis (see chapter 1) and laid the groundwork for the Watson-Crick model of DNA structure. Chromatography is probably the most sensitive method for separating both small and large molecules. By appropriate choice of the chromatographing material and the eluting solvent, separations can be achieved based upon differences in charge, solubility, molecular weight, partition coefficient, specific binding, and combinations of these factors. New substances have been detected (for example, the minor bases of tRNA), enzymes have been purified, micro columns (<2 cm long) have been useful in enzyme assays, and giant columns (10 meters high) have been used in massive purifications. *Gel chromatography* has been used in a quantitative way to determine molecular weights, binding constants, and to monitor exchange reactions. For example, the making and breaking (breathing) of hydrogen bonds in DNA was first detected by gel chromatography.

The invention of the *electron microscope* had an impact on chemistry, physics, and all of biology. In biology it afforded the opportunity to view biological structures that were thought to be impossible to observe and were undreamed of, in many cases. As the resolving power of the microscope was improved, direct visualization of large molecules became possible. Cecil Hall worked out techniques for adsorbing protein molecules to the very smooth surface of freshly cleaved mica and then coating the molecule with a heavy metal (in order to make the molecule visible). This development took the field of protein structure a step forward. The *negative contrast method,* in which particles are placed in a liquid opaque to electrons (usually a solution of either phosphotungstic acid or uranyl acetate), so that one obtains an image of the surface of the particle, enables structural biologists to determine with little ambiguity the structure of viruses, phages, and ribosomes. Albrecht Kleinschmidt's outstandingly simple procedure for visualizing DNA adsorbed to a film of protein was of far-reaching importance in the study of nucleic acid. Determination of molecular weight became a routine procedure; sites of protein binding were identified; base sequence homology was detected by hybridization or the *heteroduplex modification;* circular, supercoiled, and catenated DNA molecules were discovered; evidence was produced for insertion sequences and transposition elements; and various forms of replicated DNA were detected.

Electrophoresis, or the movement of molecules in an electric field, has yet to attain its potential usefulness as a laboratory technique, primarily because of great difficulties (which are still not overcome) in working out complications in the theory of electrophoresis. At first the prime importance of the technique was as a means of detecting heterogeneity of proteins. The development of electrophoresis in starch gels and of more sensitive procedures for detecting proteins within gel material led to the discovery of *isozymes* (different proteins or multiple forms of a single protein each having the same enzymatic activity). The introduction of polyacrylamide as a support enabled Klaus Weber and Mary Osborne to introduce sodium dodecyl sulfate (SDS) into the gel; this method, **SDS-gel electrophoresis,** is now a standard method for measuring the molecular weight of single-chain protein molecules. Polyacrylamide also gives better separation of RNA molecules than zonal sedimentation and the introduction of this support led to rapid advances in the study of mRNA. Later, the support material agarose became available, and electrophoresis of DNA was finally possible. A significant change in the technology used to study DNA followed the isolation of the *restriction endonucleases,* enzymes that can cleave DNA molecules at a small number of unique sites. Once these were used with agarose gel electrophoresis, physical mapping of DNA became possible, and invaluable gene maps of organisms that do not engage in genetic recombination were obtained. Two techniques of extraordinary consequence have come about as a result of this combination of restriction endonucleases and gel electrophoresis—the ability to recombine DNA fragments to create new organisms (the recombinant DNA technology) and the brilliant procedure of Allan Maxam and Walter Gilbert for determining the base sequence of DNA molecules (see problem 4-71 in chapter 4).

On occasion a technical advance has created quite unexpected possibilities far removed from the initial reason for its development. One example of this is *membrane filtration.* Nitrocellulose and cellulose acetate filters have smaller pores than conventional paper filters and are extremely useful in removing bacteria from liquids and for collecting minute precipitates such as those encountered in radiochemical assays. It was unexpected, though, that nitrocellulose would, under certain ionic conditions, bind single-stranded DNA and protein molecules as efficiently as it does. As a consequence of this binding, there resulted an extremely sensitive procedure for assaying a protein whose only known activity is to bind to DNA. In this procedure, radioactive double-stranded DNA is added to a mixture possibly containing the protein in question and the mixture is passed through a nitrocellulose filter. Although double-stranded DNA fails to bind to the filter, if a DNA-binding protein is in the mixture, the DNA will also be retained on the filter by first binding to the protein; the protein can adhere to the filter even when complexed with DNA. The ability of single-stranded DNA but not RNA to bind to nitrocellulose enables one to detect RNA complementary to a particular DNA molecule. That is, radioactive RNA (mRNA, for example), when subjected to hybridization conditions, is retained on a filter that already contains complementary single-stranded DNA. This simple procedure is the basis of all studies of the regulation of synthesis of mRNA.

To this list of techniques could also be added many others that became important once a particular instrument or piece of equipment was developed: examples are the high-speed computer, without which X-ray diffraction analysis would not be effective in determining the structure of globular proteins and other complex molecules, and the spectrometers used in circular dichroism and Fourier transform nuclear magnetic resonance spectroscopy—both spectrometers can yield extraordinary detail about molecular structure. These examples are just a few of many that might be mentioned.

The references listed at the end of this introduction contain full explanations of the techniques mentioned above. An entire book would be required for comprehensive treatment of the material of this chapter, and such a treatment will not be attempted. Instead, the problems in the chapter have been chosen to provide practice in certain techniques.

REFERENCES

GENERAL REFERENCES DEALING WITH NUMEROUS TECHNIQUES

Brewer, J. M., A. J. Pesce, and R. B. Ashworth. 1974. *Experimental Techniques in Biochemistry*. Englewood Cliffs, N.J., Prentice-Hall.

Freifelder, D. 1976. *Physical Biochemistry*. San Francisco. W. H. Freeman and Company. Some of the problems in this chapter are taken from this book.

Lehninger, A. L. 1975. *Biochemistry*. New York. Worth Publishers, Inc. Second Edition. Chapter 7.

SPECIFIC REFERENCES

Davis, R. W., M. Simon, and N. G. Davidson. 1971. "Electron microscopic heteroduplex methods for mapping sequence homology in nucleic acids." *In* Grossman, L., and H. K. Moldave, eds. *Methods in Enzymology. Vol. 21: Nucleic Acids, Part D.* New York, Academic Press. pp. 413–428. This is the most detailed description of how to do the Kleinschmidt procedure.

Fisher, L. 1969. "An introduction to gel chromatography." *In* Work, T. S., and E. Work, eds. *Laboratory Techniques in Biochemistry and Molecular Biology.* New York. American Elsevier.

Kleinschmidt, A. K. 1968. "Monolayer techniques in electron microscopy of nucleic acid molecules." *In* Grossman, L., and K. Moldave, eds. *Methods in Enzymology.* Vol. 12, Part B. pp. 361–377.

Kobayashi, Y., and D. U. Maudsley. 1969. "Practical aspects of scintillation counting." *In* Glick, D., ed. *Methods of Biochemical Analysis. Vol. 17.* New York. John Wiley & Sons (Interscience). pp. 55–153.

Maurer, H. R. 1971. *Disc Electrophoresis.* New York. Walter de Gruyter, Inc.

Oliver, R. M. 1973. "Negative stain electron microscopy of protein molecules." *In* Hirs, C. H. W., and S. N. Timasheff, eds. *Methods in Enzymology. Vol. 27: Enzyme Structure, Part D.* New York. Academic Press. pp. 616–672.

Peterson, E. A. 1964. "Acrylamide gel electrophoresis." *Ann. N.Y. Acad. Sci.,* **121:** 350–365.

Slayter, E. M. 1970. *Optical Methods in Biology.* New York. John Wiley & Sons. A superb book on electron microscopy, spectrophotometry, and other optical procedures.

Weber, K., and M. Osborn. 1969. "The reliability of molecular weights determined by SDS-polyacrylamide gel electrophoresis." *J. Biol. Chem.,* **24:** 4406–4412. The definitive paper on the subject.

Problems

○14-1. In the Kleinschmidt technique for visualizing DNA, explain why it is necessary to shadow the DNA with metal at a very low angle. Is the metal being deposited directly onto the DNA molecule?

○14-2. Explain the principle of negative contrast. The usual reagent used in this procedure is phosphotungstic acid. What is actually absorbing the electrons? How would the appearance of an empty phage head compare to that of a head filled with DNA?

14-3. (a) Explain how shadowing a phage sample allows one to measure the dimensions of the particle.

(b) A spherical virus is mixed with polystyrene spheres, 705 Å in diameter. After shadowing with tungsten, the length of the shadow of the polystyrene spheres is 1,250 Å and that of the virus 820 Å. What is the diameter of the virus? Some of the viruses have shadows ranging from 150 to 200 Å and look somewhat fuzzy. What are these?

14-4. Support films for electron microscopy are always very thin layers of either plastic or pure carbon. These films are very fragile and frequently break. Stronger films could be made of metals such as chromium. Why would such metal films not be useful?

14-5. In a DNA fraction isolated from phage-infected cells, it is suspected that there will be linked circular molecules *(catenanes)*. In electron micrographs of this DNA, 40 percent of the observed circles appear to overlap as linked circles might. What criterion should be used to ensure that the circles are linked and not just unit-sized circles resting one on top of another on the support film? How can one minimize this kind of accidental overlap?

14-6. Repressors are often analyzed by mixing them with radioactive DNA, which possesses an appropriate operator, and passing the mixture through a nitrocellulose filter. The amount of bound radioactivity is a measure of the amount of repressor.
(a) What is the physical principle underlying this technique?
(b) What control must be done in order to obtain accurate values for the amount of repressor?

14-7. Suppose that an enzyme is dissociated into four identical subunits and that you want to test for the enzymatic activity of the individual subunits. However, you must be sure that there are no tetramers remaining in the sample.
(a) What chromatographic procedure would you choose to free the monomers from the tetramers?
(b) How would you know where the tetramer would be if it were present?

14-8. In preparing hybrid DNA by renaturation of denatured DNA, there is usually some unrenatured DNA. Describe several procedures that could be used to eliminate the unwanted DNA.

14-9. A DNA molecule consisting of a 6×10^6 dalton double-stranded piece and a single-stranded extension equal in length to the double-stranded segment is adsorbed to a hydroxyapatite column. Given the fact that there is very little dependence of elution behavior on molecular weight in this range, if the column were eluted with a gradient of increasing concentration of phosphate, where would you expect this molecule to elute—in the single-stranded, or double-stranded region, or elsewhere? (See Figure 14-1.)

•14-10. An enzyme is known to require a high concentration of Mg^{2+} for activity. If the Mg^{2+} is removed, the protein is irreversibly dena-

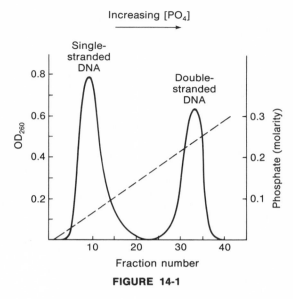

Increasing [PO₄]

FIGURE 14-1

tured. Suppose that, in establishing a purification scheme, you try both ion exchange and gel chromatography and that, in both cases, the enzyme loses activity. Explain why that might happen. In view of your explanation, what modifications could you make to improve the situation?

14-11. Electrophoresis in gels is now a common technique in molecular biology. Why is it necessary that the electrophoresis be done in solutions having low salt concentration?

○14-12. In two-dimensional chromatography and electrophoresis, such as is employed in the fingerprinting technique, does it matter which is done first? Explain.

14-13. A mixture of proteins is subjected to electrophoresis in three polyacrylamide gels, each having a different pH value. In each gel five bands are seen.
(a) Can one reasonably conclude that there are only five proteins in the mixture? Explain.
(b) Would the conclusion be different if a mixture of DNA fragments was being studied?

14-14. An enzyme has been extensively purified. By a variety of criteria it is thought to be pure—that is, it shows a single peak after chromatography in several solvents, electrophoresis at several pH values, and after centrifugation in solutions of different ionic strength and composition. When subjected to sodium dodecyl sulfate-gel (SDS-gel) electrophoresis, two bands result, one twice the area of the other. What information does this give about the protein?

Because purity is always difficult to prove, how could you prove that your hypothesis is correct? *Hint:* Use gel chromatography.

14-15. A virus contains 256 proteins, 64 having a molecular weight of 1,800 and 192 having a molecular weight of 26,000. If the virus were disrupted and analyzed by SDS-gel electrophoresis, what would be the relative distances migrated and the relative areas of the bands?

14-16. Suppose that you have isolated a protein that seems to have two enzymatic activities. This makes you suspect that you may have two proteins that copurify. To check this, you subject the preparation to electrophoresis in a polyacrylamide gel at a variety of pH values. In each gel, a single band results, but the band is sufficiently broad that you suspect that it is really two bands which do not resolve. In an SDS gel, a single, broad band is also found. Because a protein with two enzymatic activities is rare, it is necessary to try a little harder to see if the breadth of the band is due to the presence of two bands. What parameters could you vary to improve resolution by electrophoresis? What other methods (nonelectrophoretic) might you try?

○14-17. When DNA is analyzed by treatment with a restriction endonuclease followed by gel electrophoresis, it is almost always necessary to heat the mixture to 65° C before layering it on the gel. If this is not done, extra bands are seen. Why is this the case?

14-18. In analyzing DNA by treatment with a restriction endonuclease followed by gel electrophoresis, it is observed that in addition to the expected bands, there is DNA present in all regions of the gel starting from the position of the most slowly-moving fragment and past that of the most rapidly-moving fragment. If the DNA is analyzed before treatment with the restriction enzyme, no such material is seen. Explain.

14-19. A DNA molecule has the restriction map shown in Figure 14-2 when digested by the EcoR1 restriction endonuclease (numbers refer to relative distance of each cut from the left end of the molecule). When analyzed by gel electrophoresis (after heating the digested sample to 65° C), instead of the expected five bands, the second pattern shown is observed. Explain.

●14-20. A phage DNA molecule has short, complementary, single-stranded ends and circularizes when the phage infects a bacterium. The restriction map and the gel band pattern obtained after treatment of a free DNA molecule with the Bam restriction endonuclease are shown at A and B in Figure 14-3. You are studying the properties of DNA isolated from an infected bacterium. Under certain conditions the pattern of fragments obtained after enzymatic digestion and gel electrophoresis is that shown in panel C. What is the structure of the DNA under these conditions?

EcoR1 restriction map

Pattern after gel electrophoresis

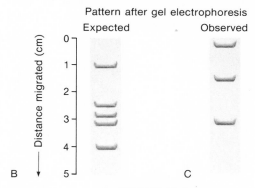

FIGURE 14-2

Bam restriction map

Pattern after gel electrophoresis

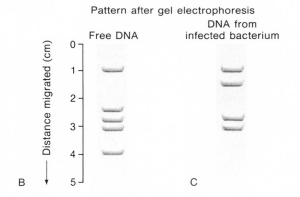

FIGURE 14-3

14-21. You are studying a protein which you believe to form dimers. You have chosen to use ultracentrifugation and to use the standard schlieren optical system of the ultracentrifuge. Under particular conditions you observe the sedimentation pattern shown below. What fraction of the molecules (in terms of the number of monomer molecules) are dimers. *Hint:* trace the pattern in Figure 14-4 on graph paper.

FIGURE 14-4

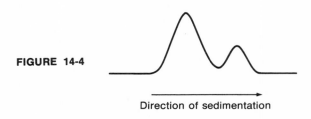

Direction of sedimentation

14-22. You are studying the effect of X-irradiation on phage B3 DNA molecules to measure the rate of production of double-strand breaks. The DNA is analyzed by ultracentrifugation using absorption optics. The sedimentation patterns shown in Figure 14-5 are observed after two X-ray doses. The molecular weight of the DNA is 20×10^6. What fraction of the molecules is broken at each dose and what is the rate of production of the breaks?

14-23. An enzyme has a sedimentation coefficient of 16 at pH 7. At pH 3 the sedimentation coefficient drops to 11 and at the same time the enzyme loses all activity (even when the pH is restored to 7). It is known that there is no change in molecular weight at pH 3.
(a) What is the approximate shape of the protein?
(b) If the first value was 4 instead and the value at pH 3 was 10, then what would you conclude?

○**14-24.** The density of DNA in CsCl is approximately 1.7 g/cm³ and that of most proteins is approximately 1.3 g/cm³. What would you expect to be the density of a typical bacteriophage which is 50 percent protein and 50 percent DNA?

○**14-25.** If the density of a typical protein is 1.300 g/cm, what would the density be if the protein contained ^{15}N instead of ^{14}N? ^{13}C instead of ^{12}C?

14-26. Many DNA molecules have long tracts of pyrimidines on one of the strands (and of course, purine tracts on the other). If the DNA is denatured in the presence of a polyribopurine and then centrifuged to equilibrium in a buoyant solution of Cs_2SO_4, the denatured DNA frequently forms two bands. It is known that the density of RNA is approximately 0.100 g/ml greater than DNA. Explain why there are two bands, what material is in each band, and

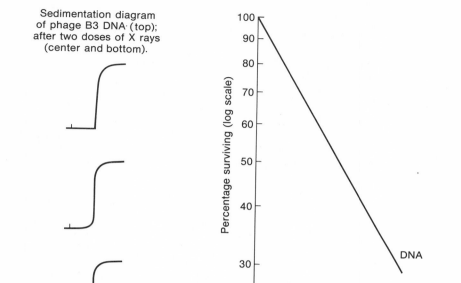

Sedimentation diagram of phage B3 DNA (top); after two doses of X rays (center and bottom).

FIGURE 14-5

whether the separation of the two bands would increase or decrease as the molecular weight of the polyribopurine increases.

●**14-27.** A virus sample is suspended in 3 M potassium tartrate and then centrifuged. After 6 hours of centrifugation it is observed that there is no pellet at the bottom of the centrifuge tube but, instead, there is a thin translucent film floating on the surface of the liquid. In another experiment the virus sample is exposed to ether and then centrifuged in the same way. When this is done, after centrifugation a pellet is found and there is no material on the surface of the liquid. Explain what is happening and what might be the composition of the virus.

○**14-28.** Which would have a higher sedimentation coefficient, a rigid rod or a flexible rod, both having the same radius and same mass?

○**14-29.** A solution of a substance at a concentration of 32 μg/ml, having a molecular weight of 423, has an absorbance of 0.27 at 540 nm measured with a 1-cm light path. What is the molar extinction coefficient at 540 nm, assuming that Beer's Law is obeyed at an absorbance of 0.27?

○**14-30.** A particular molecule has a molar extinction coefficient of 348 at 482 nm. A solution of this molecule has an $OD_{482} = 1.6$. When diluted to $1:1$, $1:2$, $1:3$, $1:4$, $1:5$, and $1:6$ concentrations, the values of OD_{482} are 1.52, 1.42, 1.05, 0.84, 0.70, and 0.61, respectively. What is the molarity of the original solution?

●14-31. A protein X is known to bind to DNA. It is of interest to measure the extent of binding as a function of salt concentration. To do this, ^{14}C-labeled protein is mixed with ^3H-labeled DNA and the mixture is sedimented through three sucrose gradients containing 0, 0.1, and 1.0 M NaCl, respectively. A sedimentation marker is added whose s is independent of salt concentration, and the centrifuge is stopped when this marker has moved through $\frac{2}{3}$ of the centrifuge tube. The results of the experiment show that there is no binding when NaCl is absent, that it is very strong in 0.1 M NaCl, and that it is relatively weak in 1.0 M NaCl. Furthermore, when the protein is bound, the DNA is in a highly compact form. Draw sucrose-gradient profiles that would lead to this conclusion.

●14-32. When a DNA molecule is heated, base pairs are disrupted, the amount of disruption increasing with temperature. At a critical temperature, the two strands separate. Draw a graph of the sedimentation coefficient, s, versus temperature. Indicate the position of T_m. Assume that the DNA molecules are in a solution of high ionic strength so that single-stranded DNA molecules are flexible. Denote the s-value of DNA at 25° C as s, and that at higher temperatures as s_T.

15

Phages

Introduction

The greater part of our understanding of the molecular basis of fundamental biological processes has come from studies of phage biology. Phages are particles consisting of a single molecule of nucleic acid packaged in a protein coat and capable of replication only within a host organism. Most known phages grow in bacteria and hence are designated *bacteriophages;* however, there are also fungal phages, and a few that grow on blue-green algae—the cyanophages. The nucleic acid is generally double-stranded DNA, but several well-known phages contain single-stranded DNA and a few contain RNA. The DNA in the phage particle is usually a linear molecule, but in a few phages it is circular. Phage particles have three basic morphologies—a polyhedron with a single tail (the most common form), a polyhedron without a tail, and a rod. Any phage strain is capable of growth on only one or a very small number of species of hosts. When a species barrier is crossed, the hosts are usually closely related—for example, some *E. coli* phage species are also capable of growth on the related enterobacteria *Shigella* and *Salmonella*. On the other hand, phage species whose host is *Pseudomonas aeruginosa* usually cannot grow on the related host *P. fluorescens*.

Phage particles are very widespread in nature, being found in animal waste, soil, and water. Furthermore, there are probably phage species capable of infecting all bacterial species. Phage DNA is even more prevalent in the bacterial world than infective phage particles themselves, in that most bacteria contain phage DNA as part of the bacterial chromosome. Under certain circumstances, a phage DNA molecule can become integrated into the DNA of its host (see "Lysogeny," later in this introduction). Initially, the ability to lysogenize may help the host survive, but in time the advantage is minimized. If the presence of the phage DNA does not confer any survival advantage and if damage to the phage DNA does not reduce the ability of a lysogenic bacterium to survive in nature, mutations will gradually accumulate in the phage DNA and most of the phage functions may be lost. What may have survived is a gene for phage tails or for a head protein, and it is through occasional appearance of gene products such as these that phage DNA is detected within many bacterial species. When such DNA is present, the bacterium is said to harbor a *defective* or *cryptic prophage.*

DETECTION OF PHAGES

A phage particle is usually detected by the formation of a *plaque,* that is, a clear area in a turbid layer of bacteria. If 10^8 bacteria are allowed to grow on an agar surface, the resulting colonies will form a confluent, highly turbid layer called a *bacterial lawn.* If a single phage particle is also on the surface at the time of addition of the bacteria, it will multiply in one bacterium and about 100 phage particles will be produced; these will infect 100 bacteria, multiply, and produce 10^4 phage particles. Because the time required for each generation of phages to mature is nearly that needed for the doubling of the number of bacteria, the number of phage particles will greatly exceed the number of bacteria in a single location and there will be a region on the agar surface devoid of bacteria. This clear region is a plaque.

Plaques can be large or small, depending (1) on the relative length of the life cycles of phage and bacteria and (2) on the size and shape of the phage. They may be clear, slightly turbid, very turbid, or possess a halo, for a variety of reasons.

BASIC LIFE CYCLE OF A BACTERIOPHAGE

The life cycle of all bacteriophages follows the same basic pattern, although differences in detail exist. The various stages of the *lytic pathway,* that is, that pathway by which infection by a single phage results in phage production and lysis of the host, are the following.

1. Adsorption. The phage binds more or less tightly to specialized receptors on the cell wall* by means of binding sites on the tail of the phage or, if there is no tail, on the surface of the phage head. The maximum number of phage particles that can adsorb to a single bacterium is equal to the number of receptors and can be as many as 5,000. In a particular experimental situation the average number of particles adsorbed per bacterium is termed the *multiplicity of infection* (moi).

2. Penetration. The tailed phages inject their DNA into the host cell through the tail. The initial penetration involves a local dissolving of the bacterial wall followed by a contraction of the phage tail. Penetration of the phage DNA probably is a result of pressure within the head. How the transfer is completed and other aspects of the penetration process are poorly understood.

3. Transcription of phage DNA and reduction of host transcription. These processes vary widely with different phages. The host RNA polymerase is responsible for the initial synthesis of phage mRNA. Then usually some phage gene product is synthesized whose function is to prevent or reduce host transcription. In some bacteria the bacterial RNA polymerase is totally inactivated and a new RNA polymerase is made which is coded for by a phage gene. In other bacteria the bacterial RNA polymerase is modified so that it no longer recognizes host promoters but efficiently transcribes from promoters on the phage DNA. A third mechanism is the synthesis of a DNase which solubilizes host DNA so that only phage DNA is present to be transcribed. In most cases, however, we do not know how the host is taken over: by one of the foregoing procedures, apparently the phage succeeds in converting the host to a phage-reproducing machine.

4. Replication of phage DNA and inhibition of the replication of host DNA. Shortly after transcription begins, gene products are made which initiate phage DNA synthesis. In phage DNA that is single-stranded, the single strand is copied by host enzymes to form a double-stranded DNA molecule, which is then transcribed. Following transcription, the double-stranded DNA replicates. Phage DNA replication uses many of the replication proteins of the host but, in all phages studied, phage-gene products were also used. Phage DNA

---

* It seems unlikely that these receptors exist for the purpose of binding phages, since once a phage adsorbs, the bacterium is usually doomed. These receptors must have other important functions. A few of these are known. For example, the receptor site for *E. coli* phage T1 is used in iron transport, and the site for male-specific phages is the sex pilus, which is required for conjugation between male and female *E. coli*.

replication occurs with the DNA in circular form whether the DNA is linear or circular in the phage particle; the only known exception is *E. coli* phage T7, whose DNA apparently remains linear. Replication is frequently but not always bidirectional. In every case, except phage T7, DNA replication proceeds in two stages—first, a circle replicates once or a few times to form two or more circles and, second, a switch to the *rolling-circle mode* of replication occurs and *concatemers* are produced. The first and second stages are called *early* and *late replication,* respectively. The mechanism for the early-to-late switch is unknown.

Inhibition of host DNA replication is a common event. For *Salmonella typhimurium* phage P22 and *E. coli* phage T4, the mechanisms seem to be simple—the host DNA is solubilized by phage-encoded nucleases. Furthermore, when a T4 particle infects *E. coli,* one of the nucleotide triphosphates required for host-DNA synthesis (but not for phage-DNA synthesis) is destroyed. However, if, by mutation, host-DNA degradation is prevented and also the precursors are supplied, host synthesis is still prevented. The mechanism of this inhibition is not known.

5. Production of phage particles.   After DNA replication has begun, a second round of transcription is initiated and, as a result, phage heads and tails, phage assembly enzymes, and a DNA-packaging system are all synthesized. For several phage species, the details of assembly of the phage head and tail and the manner of attachment of the two to form a complete phage particle are well known. However, the mechanism for cutting concatemers to form monomeric phage DNA is poorly understood, except in *E. coli* phage λ, the cutting enzyme of which has been partially purified. The manner in which DNA is folded and packaged in a phage head is still a mystery.

6. Release of phage particles.   In every phage species known, except one, an enzyme is made which dissolves part of the bacterial cell wall and thereby causes the cell to burst or *lyse.* Such an enzyme is called either a *lysozyme* or an *endolysin.* In the case of *E. coli* phage M13, the rodlike phage particle is somehow extruded through the cell wall and lysis occurs only rarely. The number of phage particles produced per bacterium is called the *burst size.*

LYSOGENY

Under certain circumstances infection of a bacterium with some phage types does not lead to phage production and lysis, but instead, the *lysogenic pathway* is followed. The phenomenon of lysogeny is best understood in *E. coli* phage λ. Following adsorption, injection, and early transcription, a repressor (also called an *immunity substance*) is made. The repressor binds to two phage DNA *operators* and prevents further transcription, so that many phage proteins, for example, heads, tails, and endolysin, are never made. Among the early proteins made before establishment of repression is the enzyme *integrase,* which catalyzes a reciprocal break-and-rejoin event between two specific sites in the phage DNA and the bacterial DNA. These sites are called *attachment sites (att).* The procedure results in insertion of the phage DNA into the bacterial DNA, and the bacterium then becomes a lysogen. Phage DNA that has become part of a lysogen is called *prophage,* and is simply part of the *E. coli* chromosome, replicating as part of it (see Figure 15-1). Synthesis of the repressor continues in the lysogen and all of its progeny. Thus, bacteria lysogenic for λ cannot be infected by a λ phage because the infected λ DNA would immediately be repressed. That is to say, *a lysogen is made immune to superinfection by a phage that has the same immunity (that is, the same repressor and operator) as the prophage.* However, phages whose immunity differs from that of the prophage can in general grow in a lysogen.

If a lysogen is irradiated with ultraviolet light or (more generally) if bacterial DNA synthesis is inhibited but RNA synthesis and protein synthesis are not inhibited, derepression occurs and phage transcription begins. Among the many λ gene-products made are two, integrase and excisionase, which together excise the prophage DNA. Under conditions of derepression a lytic life-cycle ensues and the lysogen ultimately lyses, releasing phage particles identical to that which first created the prophage.

At the time of initial bacterial infection, all known lysogenic phage species make a choice between the lytic and lysogenic pathways. Lysogenesis is the more probable outcome when a bacterium is infected by many phage particles and when the growth rate of the bacterial cells is decreasing (that is, when a bacterial culture is in the late logarithmic stage of growth). This is not improbable from an evolutionary point of view. When the number of bacteria exceeds the number of phage particles, it is advantageous for the phage population to follow the lytic pathway because the number of phages will then increase rapidly—for example, by 50-fold per bacterium life-cycle. However, if the number of phages exceeds the number of bacteria and if bacterial growth is slowing down, many phages will adsorb and only a few or perhaps no phages would be produced: thus, as bacterial growth tapers off, the number of phages might decrease. If lysogenization occurs, the phage DNA is maintained as long as the bacteria can grow. If at a later time the lysogen becomes unhealthy, the phage DNA should have a means of surviving the lysogen (of being excised); otherwise it might be lost forever.

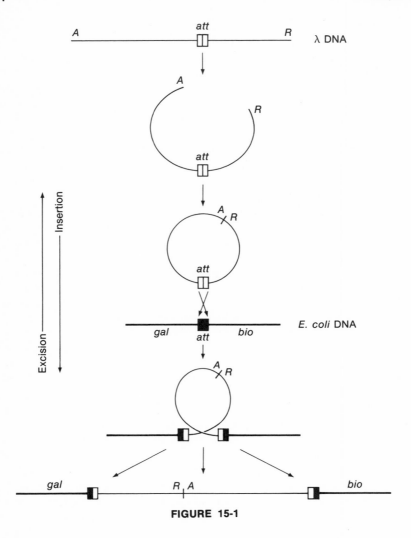

FIGURE 15-1

How the choice between the lytic and lysogenic pathways is made at the time of initial infection is only now becoming clear. The determining factor seems to be the rate of initiation of DNA synthesis. The tentative explanation is that if DNA synthesis occurs rapidly, at all times there is more DNA (actually more operators) than can be repressed by the newly synthesized repressor. Thus transcription remains turned on. On the contrary, if DNA synthesis is delayed, which appears to happen when a large number of phages infect cells in late logarithmic-stage growth, all operators are rapidly filled by repressors, transcription is turned off, and the lysogenic cycle ensues.

TRANSDUCTION

Transducing phages are phage particles that contain small regions of chromosomal DNA obtained from a *donor bacterium;* these particles can therefore carry genetic information from a donor to a recipient bacterium. Frequently, transducing phages are defective in the sense that the bacterial DNA has replaced essential phage genes (and sometimes replaces all of the phage genes).

There are two types of transducing phages, **generalized** and **specialized.** A population of generalized transducing-phage particles can transduce *any* gene carried by the donor bacterium (by donor is meant the host used for growth of the phage). These phage particles arise when the phage-maturation system packages fragments of the host chromosome—the chromosome having been degraded by phage genes during phage takeover of the host. *E. coli* phage P1 and *Salmonella* phage P22 are examples of phages whose lysates usually contain transducing-phage particles of this type.

Specialized transducing-phage particles carry bacterial DNA *from only one or two regions of the chromosome.* Furthermore, specialized transducing particles arise only when phage lysates are prepared by inducing lysogens. Why this is so can be seen in Figure 15-2, which indicates how the transducing-phage particles arise in populations of *E. coli* phage λ. Aberrant excision of the prophage produces λ transducing-phage particles containing DNA from either the *E. coli* galactose *(gal)* or biotin *(bio)* operons. When the galactose genes are picked up, phage tail genes are lost and the transducing-phage particle is defective—the notation λ*dgal* (d for defective) indicates such a phage. On the contrary, the biotin transducing-phage particle usually lacks only the nonessential genes *int, xis, red,* and *cIII,* and therefore are viable; they are called λ*pbio* (p for plaque former). Occasionally a biotin transducer will contain a larger piece of chromosomal DNA, thus lacking the essential *N* gene; this would be a λ*dbio.* If an essential region of the genetic map of the phage extending from genes *A* through *F* are replaced by a hypothetical bacterial gene *bac,* the notation λ*dbac* (Δ*A–F*) is used to indicate the genotype.

HOST MODIFICATION AND RESTRICTION

When phages infect a bacterial population of type A, most of the phage particles produced are viable and will form plaques on A. However, often if a different bacterium of the same species but of type B is used to form a lawn, only one phage in 10^6 will make a plaque. If the phage particles in one of these rare plaques are isolated and tested for

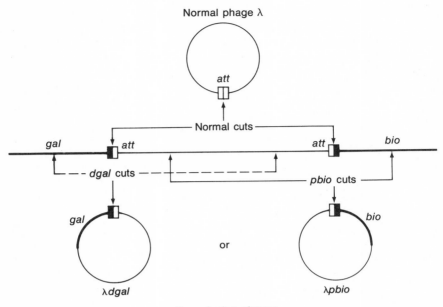

Transducing phages

FIGURE 15-2

plaque-forming ability on B, most, if not all, are viable. These phage
particles may or may not form plaques on A. This phenomenon is
called host modification and restriction. The phage P grown on
bacterium A is said to be **modified** by A and is denoted P · A. These
phages are **restricted** by B and the rare plaques on B resulting from a
P · A phage that escapes restriction are said to contain P · B phage. If
P · B fails to grow on A, then A both restricts and modifies. The
molecular explanation for the phenomenon is the following. Bacteria
contain enzymes called **restriction endonucleases,** which recognize
certain base sequences and cut DNA at these sites. Identical sequences may be present in the bacterial DNA itself; to avoid self-
destruction, the bacteria contain sequence-specific adenine and
cytosine methylases which methylate bases in the restrictable sequence and render it resistant to the endonucleases. Therefore P · A
contains methylated A sequences but nonmethylated B sequences and
is susceptible to attack by the B endonuclease. The rare phage P · B
which has escaped restriction will then have a methylated B sequence, but not a methylated A sequence, and is therefore resistant to
attack by the B enzyme. The pattern of restriction and modification
need not be symmetric since, if the ability to restrict and modify is
denoted by the genotypes r^+ and m^+ respectively, a bacterium might

have any of the genotypes r^+m^+, r^-m^+, and r^-m^-. Note that the genotype r^+m^- is not a viable combination since such a bacterium would be self-restrictive.

HOW DO BACTERIA SURVIVE ALL OF THE PHAGES IN NATURE?

With the large number of phages in the world it would seem that in time all bacteria would be infected and destroyed. There are two reasons why this does not happen. First, bacteria have evolved protective mechanisms, one example of which is the restriction system just described. A more general phenomenon is that in the course of time the phage receptors in the cell surface have changed by mutation so that they can be recognized by only a very small number of phage types; in this way, adsorption is blocked. Evidence for this comes from two observations: *E. coli* phage T4 cannot adsorb to the bacteria of the genus *Aerobacter*, yet if T4 DNA is introduced into *Aerobacter* cells through *in vitro* manipulation, normal T4 phages are made; and if a population of an *E. coli* strain (call it Q) is infected at a high multiplicity of infection by phage P, resistant cells denoted *E. coli* Q/P ("Q-bar-P") are found at a frequency of about one per 10^6–10^7 cells, and these are unable to adsorb P.

Second, in accord with the Law of Mass Action, the probability of phage adsorption is proportional to the product of phage and bacterial concentrations. In the laboratory, if both concentrations are 10^9 per ml, there is approximately 90 percent adsorption in 10 minutes. Thus, in nature, where bacterial concentrations of 10^5–10^6 cell/ml are typical and phage concentrations are even lower, the probability of successful infection is so low that most bacteria survive.

REFERENCES

Adams, M. 1959. *Bacteriophages*. New York. John Wiley & Sons (Interscience). Now outdated; yet it contains elementary information rarely found in modern texts.

Davis, B. D., R. Dulbecco, H. N. Eisen, H. S. Ginsberg, and W. E. Wood. 1973. *Microbiology*. New York. Harper & Row. Chapters 45 and 46.

Hayes, W. 1968. *The Genetics of Bacteria and Their Viruses*. New York. John Wiley & Sons.

Hershey, A. D. 1977. *The Bacteriophage* λ. Cold Spring Harbor Laboratory. An advanced, encyclopedic collection of review articles and research papers.

Stent, G. 1963. *Molecular Biology of Bacterial Viruses*. San Francisco. W. H. Freeman and Company. Outdated but still very informative; interesting historical approach.

Stent, G., and R. Calendar. 1978. *Molecular Genetics.* San Francisco. W. H. Freeman and Company. Chapters 10, 13, and 17. Excellent elementary description of phage biology including details of phage genetics not found in other texts.

Watson, J. D. 1976. *Molecular Biology of the Gene.* Menlo Park, Calif. W. A. Benjamin. Chapters 7 and 15.

Zinder, N. 1975. *RNA Phages.* Cold Spring Harbor, N.Y. Cold Spring Harbor Laboratory. The most complete collection of information about the biology of the RNA phages.

Problems

○15-1. Give several examples of mechanisms by which a phage converts a bacterium to a phage-producing machine.

15-2. Phage T4 normally forms small clear plaques on a lawn of *E. coli* strain B. A mutant of *E. coli* called B/4 is unable to adsorb T4 phage particles so that no plaques are formed. T4*h* is a host-range mutant phage capable of adsorbing to *E. coli* B and to B/4, and forms normal-looking plaques. If *E. coli* B and the mutant B/4 are mixed in equal proportions and used to generate a lawn, what will be the appearance of plaques made by T4 and T4*h*?

○15-3. Some types of phages produce small plaques and some produce large plaques.
(a) Give several reasons why this might be so and explain why populations of smaller phages tend to produce larger plaques.
(b) Even for a given type of phage there will be varying plaque size if phage particles are added to agar containing bacteria. However, if phages and bacteria are incubated together for several minutes before adding agar, the plaques are more uniform in size. Propose an explanation.
(c) Some phages produce clear plaques with large turbid halos. These halos tend to increase in size even after phage production has stopped. Furthermore, very few phage are recovered from the halo region. Suggest a cause for the halo.

15-4. If a culture of *E. coli* is infected with T4, the burst size is about 300, whereas with T7 it is about 100. For each, the lysis time is about 22 minutes at 37° C. However, T7 makes plaques whose diameter is roughly four times as great as T4 plaques. Propose an explanation.

15-5. Typically, a bacterium infected with a virulent phage produces from 50 to 500 phages per cell. If a phage mutant could produce only one new phage particle per cell, clearly a plaque could not be produced. Normally 10^8 bacteria are put on an agar surface when assaying for phages by the plaque method. The maximum bacterial concentration reached is about 10^{10} bacteria per Petri dish. Assuming that the doubling-time of the bacterium and the length of the phage life cycle are the same, what is the minimum burst size that a phage mutant can make and still produce a visible plaque? The smallest visible plaque contains about 10^5 phages.

○**15-6.** Consider a phage with the strange property that productive infection occurs only if at least two phages adsorb to the bacterium. If you have 10^8 bacteria and add to this culture 3×10^8 phages, how many bacteria will be productively infected?

15-7. The number of phages in a lysate is being determined by plating. First 0.1 ml of the lysate is diluted in 10 ml of buffer (tube A). Two successive dilutions of 0.1 ml to 10 ml are made (tubes B and C), followed by two successive dilutions of 1 ml to 9 ml (tubes D and E). Then, 0.1 ml from various dilution tubes is placed in agar with sensitive bacteria, and the next day, plaques are counted. The number of plaques obtained from the 0.1 ml aliquot from tubes C, D, and E are 2,010, 352, and 18, respectively. What is the best estimate of the number of phages per ml in the lysate? Why do the number of plaques per dilution tube not simply follow the dilution?

○**15-8.** One ml of a bacterial culture at 5×10^8 cells/ml is infected with 10^9 phages. After sufficient time for >99 percent adsorption, phage antiserum is added to inactivate all unadsorbed phages. The infected cell is mixed with indicator cells in soft agar, and plaques are allowed to form. If 200 cells are put on a Petri dish, how many plaques will be found?

15-9. Suppose a culture is to be simultaneously infected with phages A and B. What ratio of phages to bacteria must be used to ensure that 90 percent of the bacteria are infected with at least one phage of each type?

15-10. An *E. coli* culture is infected with T2 and T2*h* (host-range mutant) at a ratio of 5 phages each per bacterium. A lysate is obtained, a portion of which is adsorbed either to (a) wild-type *E. coli* or (b) *E. coli* B/2 (unable to adsorb T2), in each case at a multiplicity of infection of 0.1. Antiserum is added to remove unadsorbed phage. The infected cells are plated on a lawn of B/2. Twice as many plaques are formed with infection (a) than with (b). Explain.

○**15-11.** Rank the *E. coli* T phages in order of increasing molecular weight of their DNA.

●**15-12.** If *E. coli* phage T4 is incubated in 4 *M* NaCl for 10 minutes and then diluted 100-fold into distilled water, the phage head ruptures and releases the DNA. There is no rupture if the phages are put in

4 M NaCl and then are immediately diluted into water. Rupturing is called **osmotic shock.** Some of the survivors of osmotic shock are mutants which are resistant to osmotic shock. Some of these resistant mutants are found to be inactivated when heated to temperatures at which the wild-type is not damaged. What part of the phage has probably been altered in these mutants? Explain.

○15-13. In an experiment in which 10^7 bacteria are mixed with 5×10^7 phages, how does one determine the actual multiplicity of infection and the number of infective centers?

●15-14. A lysate of phage R is found to have the following peculiar property. By plating on sensitive bacteria it is found that there are 10^8 plaque-forming units per ml. If 0.1 ml of R is mixed with 10^8 bacteria, there are 10^7 infectious centers. However, if the bacteria are infected at moi = 3 with a mutant p^-, and then superinfected with 0.1 ml of the R lysate described above, 3×10^7 infectious centers result. If the bacteria are infected at moi = 3 with a mutant q^-, and then superinfected with 0.1 ml of the R lysate, 10^7 infectious centers are found. Explain.

○15-15. Define the following terms: infectious center; multiplicity reactivation; eclipse phase; marker rescue; lysis-from-without; heterogenote; phenotypic mixing; lysis inhibition; and phage ghost. Give examples where possible.

15-16. What would you expect to happen if, after virulent phage infection (such as with T4), all the phage genes were transcribed and translated at once following phage DNA injection?

●15-17. In a broth consisting of glucose and yeast extract, *E. coli* grows with a doubling-time of 30 min. Coliphage T7 has a latent period of 20 min under these conditions, and a burst size of 200 phage per infected cell. If a culture of 2×10^7 *E. coli* per ml is growing exponentially and 5,000 plaque forming units per ml of T7 phage are added, when will the culture lyse? A few minutes before visible lysis a certain number of bacteria will remain uninfected. The estimate for uninfected cells a few minutes before the time of lysis would be difficult to confirm experimentally, since plating of cells and phage together might lead to infection and killing of all the surviving bacteria. In practice, about 100 colony-forming units per ml are found. Some of the colonies are large and round. What might these colonies represent?

In working with this problem assume that phage adsorption is instantaneous and that multiply-infected bacteria give the same burst as singly-infected bacteria.

●15-18. A new protein X appears in infected cells. Describe various experiments that you might use to prove that the gene coding for X is located on the phage genome and is not coded for by a host gene.

●15-19. How would you show whether or not a phage-coded gene product is required throughout the infectious cycle or only at a unique time?

○**15-20.** What is meant by sup^+ and sup^- hosts with respect to phage mutant growth?

●**15-21.** P2 and P4 are temperate bacteriophages of *E. coli*. They have the following properties: (1) When a P2 phage infects a nonlysogenic bacterium, the bacterium usually bursts, giving about 100 new P2 phages. (2) When a P4 phage infects a nonlysogenic bacterium, the bacterium survives, but becomes lysogenic for P4. (3) When P2 phage and P4 phage coinfect the same bacterium, lysis of the bacterium gives 100 P4 progeny and no P2 progeny.

If 3×10^8 P2 phages and 2×10^8 P4 phages are added to 10^8 nonlysogenic bacteria, then:

(a) How many bacteria will not be infected at all? (These are bacteria to which no phage adsorb.)

(b) How many bacteria will become lysogenic for P4? (Bacteria to which P4 adsorb, but no P2 adsorb.)

(c) How many bacteria will produce P2 progeny? (Those to which P2 adsorb, but no P4 adsorb.)

(d) How many bacteria will produce P4 progeny? (Bacteria which are infected by at least one P2 and one P4.)

15-22. A 10 ml broth culture of *E. coli* at 2×10^7 colony-forming units per ml is infected with 5×10^5 plaque forming units of virulent phage T5, at 37° C. Under these conditions *E. coli* divides every 30 minutes. Also every 30 minutes T5 gives a burst of 150 phages per cell. If the phage adsorb without any significant delay (1 minute) when will the culture lyse visibly?

●**15-23.** It is known that phages cannot multiply in nongrowing bacteria. However, if 0.1 ml of a stationary *E. coli* culture at 2×10^9 cells/ml is mixed with 0.1 ml of a T4 phage suspension at 10^{12} phages/ml, lysis occurs shortly afterward. This is called ***lysis-from-without***. It is known that purified phages are much less effective at lysis-from-without than the original lysate and that a lysate from which the phages have been removed is virtually ineffective.

(a) Explain the cause of the phenomenon.

(b) Would the unpurified lysate be able to lyse a culture of a T4-resistant mutant of *E. coli?*

(c) Could the phage-free lysate enhance the effectiveness of lysis-from-without by a purified sample of phage T7?

15-24. If an *E. coli* culture is simultaneously infected by phages T4 and T7 each at a multiplicity of infection of 5, only T4 phage will be produced. From what you know about T4 biology, propose a simple explanation. What would you expect to happen if the T4 was added 10 minutes after addition of T7?

○**15-25.** What is the function of the T4 dCTPase (deoxycytidine triphosphatase)?

○**15-26.** Describe the course of T4-phage DNA synthesis following infection of *E. coli* with a T4 mutant which cannot synthesize (a) cytidine hydroxymethylase, (b) α-glycosyl transferase.

○**15-27.** In an infection with T4 phage, which of the following are true (several answers are):

(a) *E. coli* DNA is degraded by a T4-phage enzyme.

(b) T4 phage DNA is replicated by *E. coli* polymerase III.

(c) Deoxycytidine triphosphate (dCTP) is converted by cytidine hydroxymethylase to dHTP, which is then glucosylated by α-glucosyl transferase, and the glucosylated dHTP is then incorporated into T4 DNA by a DNA polymerase.

(d) Early mRNA is made by using *E. coli* RNA polymerase with an *E. coli* σ factor.

(e) Late mRNA synthesis is delayed until T4 DNA has replicated.

○**15-28.** If a T4 phage has a large deletion, which of the following will be true?

(a) The activity of some protein will be greatly altered or altogether missing.

(b) The phage DNA will be smaller.

(c) The terminal redundancy will be larger.

(d) The phage DNA will be the same size.

(e) Cyclic permutation will be eliminated.

15-29. Suppose you had a phage whose linear DNA is synthesized by the rolling circle mode* and is packaged by "the headful rule" (that is, DNA is added to a head of fixed size until no more DNA can fit). However, the DNA is normally neither terminally redundant nor cyclically permuted. You find a mutant strain of this phage, the DNA of which has a deletion in a nonessential gene. This phage is used to infect a bacterium, and many phage are produced. The DNA is isolated and is treated with an exonuclease, which removes a few bases from the 5′-P end. The treated DNA is then exposed to conditions which could circularize T4 DNA (if it were also pretreated with the exonuclease). This DNA is examined by the electron microscope. Would circles be found?

15-30. After infection by the virulent phage T4, all host DNA synthesis is rapidly shut off even in the absence of degradation of host DNA. You decide to study this phenomenon.

(a) List three possibilities (as unrelated as possible) for how the shutting off might be accomplished (but remember that T4 has to grow).

(b) How would you try to test for these possibilities *in vitro*? (Try to think of experiments that are simple and yet serve to define the mechanism.)

15-31. How would you isolate a T4 mutant that

(a) Is defective in thymidine kinase?

(b) Is mutant in gene 30 (ligase)?

(c) Is defective in inducing host DNA degradation?

---

* Do not concern yourself with how a linear DNA could get to the rolling circle mode—let it be by magic.

15-32. (a) Could you determine a unique origin of replication of T4 DNA by making use of the technique of partial denaturation mapping? (b) Can one discriminate whether early or late T4 genes are injected first into the *E. coli* host cell? Explain.

○15-33. List some T4 gene products that would be expected to act catalytically; list those products that would be required in stoichiometric amounts.

15-34. *E. coli* phage T4 synthesizes many enzymes, for examples, thymidylate synthetase and deoxycytidine deaminase, which are not essential for growth. Explain how such genes may have evolved as part of the T4 genome.

●15-35. How would you show whether a group of genes is organized in the form of an operon on the T4 genome? If phage operons did exist, how might nonsense mutants affect complementation?

15-36. Some *E. coli* phages code for an enzyme SAMase which cleaves S-adenosylmethionine. Propose a possible function for this enzyme.

15-37. Why are 30 phage T7 proteins observed on gels prepared from T7-infected cells, but genetic analysis using amber mutants reveals only about 19 complementation groups and genes?

15-38. Is all of the coding capacity of the T7 genome expressed? How would you determine how much is expressed?

●15-39. How would you isolate a mutant in which T7 ligase is deleted? Write out a protocol.

●15-40. How does one prevent the formation of phosphorylated proteins during ^{32}P labeling of RNA in T7-infected cells? If RNA and proteins do get labeled with ^{32}P after T7 infection, how does one detect and identify each species?

●15-41. *E. coli* phage T7 codes for a protein kinase which phosphorylates proteins, in particular, the *E. coli* RNA polymerase, which is thereby inactivated.
(a) What might be the function of such an enzyme?
(b) What is the earliest stage of the infection at which this enzyme might function?
(c) The enzyme is apparently not essential, since mutants lacking the enzyme are still viable. What characteristics might you expect such a mutant to have?
(d) Phage T4 lacks this enzyme. How does it accomplish the functions mentioned in (a)?

15-42. Why is it possible to obtain species of well-defined molecular weight of T7-specific mRNA from infected cells?

15-43. What are some differences in the synthesis and nature of the early and late T7 phage-specific mRNA molecules?

15-44. What are some host functions that are specifically required for, and that inhibit, T7 infection?

○15-45. The *cI* gene of λ codes for the immunity repressor. If the λ is *cI⁻*, the plaque is uniformly transparent. If it is *cI⁺*, the plaque has a turbid center. Explain the plaque morphology. Would you expect T2 plaques ever to have turbid centers?

○15-46. A normal λ phage cannot successfully infect a bacterium (that is, cannot make more phage and lyse the bacterium) lysogenic for normal λ.
(a) Why not?
(b) Suppose the phage added to the lysogen is a mutant in that the left (o_L) and right (o_R) operators are mutated (that is, they are defective in the same way a Lac operon o^c is mutant). Can this phage successfully infect a wild-type lysogen? Explain briefly.

○15-47. If a λ phage has a large deletion, which of the following will be true? (Two answers are true. Compare Problem 15-47 and Problem 15-28 and explain differences between the two phages that would account for the different answers.)
(a) The activity of some protein will be greatly altered or altogether missing.
(b) The phage DNA will be smaller.
(c) The terminal redundancy will be larger.
(d) The phage DNA will be the same size.
(e) Cyclic permutation will be eliminated.

●15-48. Suppose you have a λ mutant that makes a clear plaque on *E. coli* strain A but a turbid one on strain B. How can this be explained?

○15-49. If one λ phage particle infects a cell lysogenic for λ, which of the following will probably happen?
(a) A normal phage cycle producing about 50 phage particles will occur.
(b) The λ DNA will circularize but will not replicate.
(c) The cell will die.
(d) The λ prophage will be excised.
(e) The λ DNA will not be injected.

15-50. If one induces a λ lysogen with ultraviolet light, the frequency of clear-plaque mutants in the resulting lysate is low, perhaps about 10^{-4}. If one simply grows a culture of a λ lysogen and analyzes the culture fluid for free phages, one finds that these are spontaneously released, occasionally, by cells in the growing culture. The frequency with which mutants in the population of spontaneously released phages form clear plaques is always higher than 10^{-4}.
(a) Why?
(b) In which of the genes *cI*, *cII*, or *cIII* do you suppose the clear mutants among spontaneously released phage reside?

○15-51. A λ phage genetic map (as obtained in standard crosses), containing only a few of the known genes, is approximated below:

A	*J*	*int*	*cI*	*P*	*Q*	*R*

Which of these genes would show the highest frequency of co-transduction with the *gal* gene if a Pl phage grown in a *gal*⁺ λ lysogenic cell were used to transduce a *gal*⁻ λ lysogen?

15-52. Refer to Figure 15-3.

(a) Does the λ*dgal* in part A of the figure show a net loss or gain of DNA as compared to the graph for normal λ (assuming the two types of particles have the same amount of protein, and the two components, DNA and protein, make independent contributions to both mass and volume of the particle?

(b) How much is the net loss of DNA in λ*dgal* No. 10, expressed as percent of the chromosome of normal λ? (Take the density of DNA as 1.7, of protein as 1.3, and of λ as 1.5.)

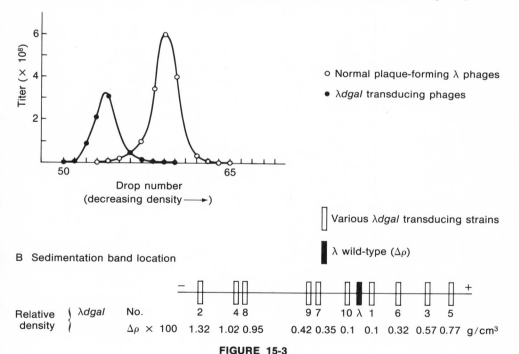

A Differences in density of normal λ and λ*dgal* transducing phages (after centrifugation).

o Normal plaque-forming λ phages

• λ*dgal* transducing phages

Various λ*dgal* transducing strains

λ wild-type (Δρ)

B Sedimentation band location

Relative density	λ*dgal*	No.	2	4 8	9 7	10 λ 1	6	3	5	
		Δρ × 100	1.32	1.02 0.95	0.42 0.35	0.1 0.1	0.32	0.57	0.77	g/cm³

FIGURE 15-3

15-53. Suppose a λ phage has infected *E. coli* and is actively replicating its DNA. No DNA has been packaged, since late mRNA has not yet been made. If a T4 phage is then added, will the λ phage still be made?

There exist certain transducing λ phages that carry the genes for
synthesizing tryptophan. The locus of the *trp*-gene insertion is
shown in Figure 15-4.

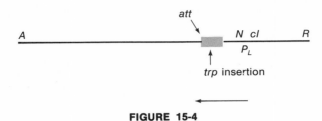

FIGURE 15-4

The tryptophan (*trp*) operon is transcribed in the direction indicated by the horizontal arrow—that is, in the same direction as the
λ phage mRNA that goes leftward from p_L. There are several
classes of tryptophan-transducing phages that differ according to
the size of the λ DNA molecule replaced by *E. coli* DNA. In each
class, the size of the inserted bacterial DNA is the same size as the
λ DNA which is absent, the bacterial DNA contains the tryptophan
synthetase gene, the insertion begins at *att* and moves to the right,
and no essential phage genes are missing.

These phages have the property that when they infect a
bacterium lacking tryptophan synthetase, this enzyme is made.
(a) If the phage carries a point mutation in the λ *N* gene, no tryptophan synthetase is made. Why not? What part of the *trp* operon
must be missing for this to be the case?
(b) Consider a transducing phage whose piece of bacterial DNA is
so large that part of the λ *N* gene is replaced by bacterial DNA.
Will any tryptophan enzymes be made? Explain briefly.

o15-55. Remember the *E. coli* gene order *gal attλ bio*, where *attλ* = *BB'*
is the attachment site in a nonlysogenic bacterium. If phage P1 is
grown on a nonlysogen, some of the transducing phages carry the
gal gene, some carry *bio*, and some carry *gal* and *bio* genes. The
ratio of phages transducing *galbio* to those transducing *gal* is 0.01.
If the P1 phage was grown instead on a λ lysogen, would you
expect the number of particles transducing *gal bio* to increase,
decrease, or stay the same? Why?

15-56. A friend has isolated a mutant of bacteriophage λ which shows
greatly elevated genetic recombination (mediated by the λ *red*
pathway). However, this "excess" recombination occurs only in
one region of λ DNA, between genes *P* and *R*. Your friend knows
something of site-specific initiation of recombination and pro-
claims that the mutant must have an "initiation-site" mutation that
gives rise to a "hot spot" where recombination can start with great
frequency. Propose two other explanations, and a way to distinguish between the possibilities.

●**15-57.** You have given a friend a mutant λ phage which is defective in gene *J* and in one of the *red* genes for phage-mediated generalized recombination. He complains that growth of the *J⁻red⁻* phage in *rec⁻* cells lysogenic for a *P⁻red⁻* prophage yields *J⁺P⁺* recombinants (the *red⁻* mutations carried by the two phages are identical). Can you explain this "extraordinary" recombination? How might you show by a genetic experiment that your explanation is probably correct?

Hint: Draw a picture of the integration reaction with the superinfecting *J⁻red⁻* DNA (assuming *PP′* × *PB′* or *PP′* × *BP′* are allowed reactions), and consider that the normal λ maturation processing applies to this situation.

15-58. Suppose you have a tandem double lysogen (that is, two prophages are next to one another) and both prophages are *int⁻* and are incapable of DNA replication. Suppose both prophages contain a mutant repressor that is temperature-sensitive, so that at high temperature mRNA can be made. If this double lysogen is heated to the high temperature, which of the following will occur?
(a) No phage will be produced.
(b) One phage will be produced per cell.
(c) Two phages will be produced per cell.
(d) *Only* transducing phage will be produced.

15-59. When Hfr males conjugate with *F⁻* cells lysogenic for λ, zygotes normally survive. However, when Hfr males lysogenic for λ conjugate with *F⁻* nonlysogens, zygotes produced from matings that have lasted for almost two hours lyse, owing to the zygotic induction of λ.
(a) How can you explain zygotic induction?
(b) How can you determine the locus of integrated λ prophage?

●**15-60.** *E. coli* Q is *recA⁻* (that is, it is deficient in bacterial recombination) and contains a portion of a λ prophage—only the *J* gene and the right prophage-attachment site (see the λ map in Appendix E, page 242). The *J* gene is mutant and codes for the *h* character. Wild-type λ cannot adsorb to the λ-resistant strain K/λ, but a λ mutant carrying the *h* marker can. If a *cI⁻* mutant of λ is plated on a bacterial lawn consisting of a 1:3 mixture of Q and K/λ, plaques result which are easily seen, although they are somewhat turbid. Many phages recovered from the plaques can form plaques on K/λ. When a mutagenized stock of λ*cI⁻* is plated on this mixture, some plaques are found which are so turbid that they are almost not visible. These plaques contain no phages that can form plaques on K/λ. What is the probable genotype of the phage producing these very turbid plaques?

●**15-61.** (a) The λ amber mutant *A32* fails to plate on *E. coli* strain 594, since this strain is *sup⁻*. The λ mutant *int6* is defective in normal prophage excision. The λ *cI857* mutation renders the *cI* repressor inactive at 42° C, whereas it has normal activity at 32° C. You have just lysogenized 594 with λ *cI857 int6 A32* and have a collection of lysogens. Design a genetic test to distinguish the single lysogens from the dilysogens. (It is known that 40 percent of all lysogens are tandem dilysogens and 60 percent are single lysogens.)

(b) On another occasion, *E. coli* strain B583 is lysogenized. This strain contains a sex factor on which the λ attachment site is located. This site is deleted from the chromosome so that a prophage can be inserted only in the sex factor. What simpler genetic test could now be used to distinguish the single lysogens from the dilysogens?

15-62. The λ mutant *ti* 12 makes small plaques and has a burst size about 20 percent that of wild-type λ. In a mixed infection of *ti* 12 and *ti*⁺, no *ti* 12 phage are produced. By use of a density label it is known that in the mixed infection *ti* 12 has failed to replicate. Furthermore, if the *ti*⁺ phage is O^-P^- (that is, mutant in the only λ genes needed for λ DNA replication), in a mixed infection with *ti* 12 O^+P^+ and *ti*⁺O^-P^-, again no *ti* 12 phage appears in the burst; the yield of *ti*⁺O^-P^- is that of the wild-type. What kind of defect is *ti* 12?

●**15-63.** A λ mutant is isolated which is defective in that, although it forms plaques normally, a lysogen containing a single mutant prophage fails to produce phage when induced. The mutant lysogenizes with normal frequency. In studying this mutant, you infect bacterial strain R with this mutant at a moi = 3. Strain R has been obtained by lysogenizing a *gal*⁻ bacterium containing a segment of a λ prophage that has only the right prophage-attachment site, with a *gal*⁺-transducing phage. The resulting structure of this bacterium is shown in Figure 15-5, where the light line represents the λ prophage.

| gal⁻ | PP′ | | gal⁺ | BB′ | bio |

FIGURE 15-5

Strain R contains a normal *cI* repressor and therefore is immune to superinfection. When R is infected with the λ mutant and the infected cells are plated on an indicator agar (galactose tetrazolium) in which *gal*⁺ and *gal*⁻ colonies are white and red, respectively, roughly equal numbers of white and red colonies are found. In what gene does the λ mutant map and what type of mutant is it?

●**15-64.** (a) Phage T4 has a circular genetic map because the population of linear DNA molecules in a lysate of phage T4 have permuted terminal sequences. Suppose you are studying a phage φ42, which has a circular genetic map, but a linear, unique DNA sequence in its phage particles. Propose a formal model (qualitative and non-mathematical) which can explain this finding.

(b) Phage λ has a linear genetic map which gives the same gene order as that determined by physical techniques to exist in DNA isolated from phage particles. A possible conceptual problem is that λ is likely to replicate and recombine as a *circular* molecule. Provide an argument (qualitative and nonmathematical) that recombination of this kind between circular DNA molecules can lead to the linear map found when we study the phage particles that come out of the mixed infection. In other words, indicate why $A^+R^- \times A^-R^+$ yields A^+R^+ more often than $N^+P^- \times N^-P^+$ yields N^+P^+. For simplicity, assume that recombination is reciprocal, so that the initial product of recombination between circles is a double-length circle. (See the λ map in Appendix E, page 242).

●15-65. You have isolated two independent mutant strains of the phage λ; both of these mutants form clear plaques. When either mutant alone infects *E. coli* at moi = 10, no lysogenic survivors can be isolated. However, when *E. coli* is coinfected with 5 phages per cell of the first mutant plus 5 phages per cell of the second mutant, then 10 percent of the cells survive the infection, and almost all of these survivors are lysogenic. When the lysogenic survivors are induced, 99 percent of the plaques produced are clear.

(a) What λ function or functions do you think these two mutants affect?

(b) What is the most likely explanation for the increased frequency of lysogenization which occurs during coinfection by the two mutants?

●15-66. (a) Phage λ infects *E. coli* at a total multiplicity of 10. An extract is made from these infected cells and the DNA binding activity is measured in an *in vitro* system. DNA binding activity is a measure of the level of the *cI* repressor in the infected cells. The data in Table 15-1 are obtained.

TABLE 15-1

Infecting phage or phages	DNA binding activity
cI^+	4.0
cI^-	0.1
$cI^- + cIII^-$	2.0
$cI^- + cII^-$	2.0
$cI^- + y^-$	0.1
$cI^+ + cI^-\,cII^-$	2.0
$cI^+ + cI^-\,y^-$	2.0

(1) What genetic approach is involved in most of the mixed infections? (2) Why is the DNA binding activity low in two of the infections? (3) How can the DNA of λimm434 be used in these experiments?

(b) The Cro protein shuts off the *cI* repressor. If a lysogen containing λN⁻O⁻cI857 is grown at 42° C, thus inactivating the temperature-sensitive *cI*857 repressor, the Cro protein is made constitutively. Explain the data given in Table 15-2, which shows the frequency of lysogeny of the *imm*λ phage when a *cro*-constitutive lysogen is infected by the indicated phage mixtures at a multiplicity of infection of five each.

TABLE 15-2

Infecting phage or phages	Percent lysogeny (immλ phage)
immλ cI⁺	1
immλ cI⁺ + imm434 cII⁻	1
immλ cI⁺ + imm434 cII⁻	1
immλ cI⁺ + imm434 cIII⁻	1
immλ cI⁺ + imm434 cI⁺	45

15-67. *Haemophilus influenzae* restriction endonuclease III (HindIII) makes cuts in λ DNA at six defined sites. These sites are designated by the percentage of the distance from the left end of the phage λ gene map (see Appendix E, page 242), as shown in Table 15-3. This cutting generates seven fragments A, B, . . . , G, respectively. Consult the λ map to determine the genes and loci contained in each fragment. These fragments can be separated by electrophoresis on agarose gels; then λ deletion mutants can be made that lack some of the fragments for which foreign DNA can then be substituted.

The λ deletion mutants and hybrid phages are made by mixing the desired fragments under conditions where the "sticky ends" (produced by the HindIII endonuclease) anneal and sealing the nicks with ligase. The resultant DNA is transfected into CaCl₂-treated *E. coli* and put onto plates. Only the DNA molecules which have all essential genes in the proper order will be able to form a plaque.

(a) Which fragments could be omitted and still allow the λ phage to grow lytically? Which could be omitted and still allow the λ phage to lysogenize a cell (but with a prophage capable of lysing the cell after induction)?

(b) When *all* the fragments are mixed together, closed with ligase, and transfection has been carried out, some of the plaques are

TABLE 15-3

Sites of cuts made by the HindIII enzyme	Percentage of distance from left end of phage λ map
1	47.0
2	51.4
3	56.3
4	76.3
5	77.3
6	91.0

turbid and some are clear. If the phages from a number of these *clear* plaques are grown up and the DNA is extracted and cut again with HindIII, it is found that all fragments are present; when these fragments are mixed together, sealed with ligase, and transfection is carried out again, some plaques are turbid and some are clear. Explain.

(c) Foreign DNA can be substituted for fragments B, C, D, and E, and often such hybrid λ phages are viable. Which of the following DNA molecules could be substituted? Explain your answer in each case with a sentence.

(1) A fragment of the *trp* operon made with EcoRI endonuclease. The base sequences recognized by the EcoRI and HindIII nucleases are, respectively,

$$\downarrow \qquad\qquad\qquad\qquad \downarrow$$
$$\underline{\text{G A A T T C}} \qquad\qquad \underline{\text{A A G C T T}}$$
$$\text{and}$$
$$\underline{\text{C T T A A G}} \qquad\qquad \underline{\text{T T C G A A}}$$
$$\uparrow \qquad\qquad\qquad\qquad \uparrow$$

(Cuts are indicated by the arrows.)
(2) A fragment of calf thymus DNA having "sticky ends" made by using terminal transferase and dATP.
(3) A fragment of the *his* operon made with HindIII endonuclease.

(d) To be able to delete fragments B, C, D, and E, what essential λ function or functions must be compensated for in those fragments needed for lytic growth? *Hint:* Look at fragment E.

●**15-68.** Consult the genetic map of phage λ in Appendix E, page 242, for the properties of the *cI*, *E*, and *J* mutations mentioned in this problem. The *cI*⁻ mutant is inactive at 42° C but active below 39°

C. The E^- and J^- mutations are amber mutations, and lysogenic bacteria used in the following experiments contain no suppressor. A strain of *E. coli* lysogenic for λcI^-E^- is thermally induced; it lyses and produces no viable phages. This is called an E^- lysate. A similar lysogen carrying λcI^-J^- is thermally induced to give a J^- lysate containing no viable phages. Mixing E^- and J^- lysates gives viable phages at levels up to 50 phages per original cell. This *in vitro* morphogenesis requires no cell debris—the debris can be removed by centrifugation at 3,000 g without reducing the complementation. The following procedure is carried out. The E^- lysate is centrifuged at 3,000 g, the supernatant is mixed with live *E. coli* carrying an amber suppressor, and the *E. coli* are then be removed by centrifugation. This treatment yields a supernatant which is devoid of activity—the complementation with the J^- lysate is abolished. However, the cells removed from the E^- lysate will complement with the J^- lysate to give plaque-forming infective centers (infected cells).

(a) What is happening in this experiment?

(b) What is the genotype of the output phage from the infective centers?

15-69. In an *E. coli* tandem dilysogen, the left and right prophages have the genotypes $cI857\ int^-P^-$ and $cI857\ int^-A^-$, respectively. The repressor of each is active at 42° C and the prophages lack a functional integrase. P^- and A^- are genotypic markers whose locations are shown on the λ genetic map (Appendix E, page 242). If the lysogen is heated to 42° C, what will be the genotype or genotypes of the phages produced? Why will more phage particles be produced than in the experiment described in Problem 15-58?

15-70. *E. coli* $groE^-$ mutants do not allow phage to form active head particles. Recently a special λ phage variant has been isolated which can grow on *E. coli* $groE^-$ mutant host strains. This phage was constructed by starting with a λ phage strain (called phage A) whose DNA has only three sites for cleavage by the EcoRI nuclease. These sites are indicated in the restriction map, shown in Figure 15-6, by vertical arrows; the distances separating the sites

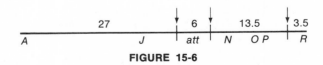

FIGURE 15-6

are given in kilobase pairs and some λ genes are indicated. The cleaved fragments are annealed at 40° C, either alone or in the presence of EcoRI nuclease fragments from *E. coli* gro^+ (wild-type) or *E. coli* $groE$ DNA. The annealed mixtures are treated with DNA ligase, and then used to transfect *E. coli gro⁺* as well as *E. coli groE⁻*. The following three mixtures give rise to plaque-forming phages, as measured on *E. coli gro⁺*: (1) λ DNA fragments alone; (2) λ DNA fragments + *E. coli gro⁺* DNA fragments; and (3) λ DNA fragments + *E. coli groE⁻* DNA fragments. Only mixture (2) gives rise to plaques which can plate on *E. coli groE⁻*; one of these plaques is picked and the unusual phages (phage B) in it are grown to high titer and the DNA is extracted, denatured, and reannealed with denatured DNA from phage A. Figure 15-7 shows the heteroduplex structure that is commonly

FIGURE 15-7

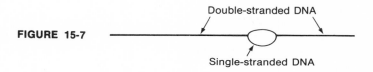

observed. After mutagenesis of phage B, it is possible to isolate some mutants which form plaques on the following bacteria: *gro⁺*, *gro⁺supF*, and *groE⁻supF*, but not *groE⁻*, where *supF* is an amber suppressor. One of these mutants was selected for study and called phage C. Phages A, B, and C were each used to infect *E. coli uvrA⁻*, which had been preirradiated with ultraviolet light, in order to abolish the capacity for host protein synthesis. Radioactive amino acids were administered during infection in order to label the phage proteins synthesized, and these proteins were separated by electrophoresis in acrylamide gels containing 1 percent sodium dodecyl sulphate, and phage protein bands were detected by autoradiography. Phage B causes synthesis of a 70,000 dalton protein (protein X), which is not made by either phage A or phage C. Phages B and C synthesize a protein of 50,000 daltons (protein Y), which is not made by phage A. Phage A synthesizes a protein of 20,000 daltons (protein Z), which is not made by phages B and C.

(a) Is the *groE* mutation dominant or recessive to the wild-type allele? Give pertinent evidence.

(b) What is the genotype of phage B?

(c) What is the genotype of phage C?

(d) What is protein X?

(e) What is protein Y?

(f) What is protein Z?

15-71. The DNA of F factors can be isolated and made free of *E. coli* DNA by banding in a CsCl gradient containing ethidium bromide. The F factor DNA is obtained as a supercoil. Individual single strands can then be purified by the following protocol. The DNA is first treated with DNase to yield an average of one nick per double-stranded molecule. The strands can then be separated, by heating in the presence of polyUG and by banding in CsCl. This complete protocol can be done with F' *gal*(λ)*bio* DNA so that the strand of F' *gal*(λ)*bio* DNA carrying only the L strand of λ DNA is obtained. If this preparation is annealed with H strands of DNA isolated from purified λ phage, what heteroduplex structures will be seen? *Note:* The DNA's of λ and F' *gal*(λ)*bio* are 17 and 82 microns long, respectively.

15-72. *E. coli* λ can follow either a lysogenic or lytic pathway. High moi favors the lysogenic pathway and low moi favors the lytic pathway. Ignoring biochemical considerations, what evolutionary pressures probably led to this effect of moi?

●15-73. Phage P22 grows on *Salmonella typhimurium.* Its DNA has a molecular weight of about 22×10^6 daltons. The genome is not unique, but consists of a circularly permuted collection of sequences, with 5 percent terminal redundancy:

$$1\ 2\ 3\ 4\ 5\ 6\ 7\ 8\ 9\ 0\ 1\ 2$$

$$3\ 4\ 5\ 6\ 7\ 8\ 9\ 0\ 1\ 2\ 3\ 4$$

$$5\ 6\ 7\ 8\ 9\ 0\ 1\ 2\ 3\ 4\ 5\ 6$$

$$7\ 8\ 9\ 0\ 1\ 2\ 3\ 4\ 5\ 6\ 7\ 8$$

$$9\ 0\ 1\ 2\ 3\ 4\ 5\ 6\ 7\ 8\ 9\ 0$$

P22 prophage is resident at a specific locus on the *Salmonella* chromosome, and the prophage genetic map is unique (not permuted). That is, P22 DNA is inserted into the *Salmonella* chromosome in one special way. P22 *int⁻* mutants have been isolated, and these mutant phages cannot insert themselves into the *Salmonella* chromosome without help (complementation) from an P22 *int⁺* phage.

Salmonella *recA⁻* mutants have been isolated, and P22 grows and lysogenizes these recombination-deficient strains normally. However, there are P22 phage mutants (P22 *erf⁻*) which cannot make plaques on *Salmonella recA⁻* host strains, although they do make plaques on *recA⁺* hosts. When these P22 mutants infect the *recA⁻* host, they cannot lysogenize, and DNA synthesis stops after one round. On the other hand, these P22 mutants can be induced from the prophage state to give a normal phage burst in *Salmonella recA⁻*:

$$recA^- \ (\text{P22 } c\text{-}ts \ erf^-) \xrightarrow{42°C} 100 \ \text{P22 } c\text{-}ts \ erf^-,$$

where *c-ts* is a temperature-sensitive mutation of the P22 repressor gene. The above strain was constructed by complementation, coinfecting *Salmonella recA⁻* with P22 *c-ts erf⁻* and P22 *c⁺erf⁺*.

(a) What do you think is wrong with P22 *erf⁻* mutants?

(b) What do you think the letters "e," "r," and "f" stand for?

●15-74. The DNA of *Salmonella* phage P22 is terminally redundant and permuted. P22 is capable of both generalized and specialized transduction. If grown on *Salmonella* carrying a resistance-transfer factor (RTF), P22 can transduce tetracycline resistance *(tet-r)* from the RTF to a sensitive *(tet-s)* *Salmonella* strain. Since P22 DNA is not large enough to carry the entire RTF DNA, the *tet-r* allele must be incorporated into the host DNA in order to be replicated. Consider a transduction in which a lysate prepared on a *tet-r* strain is used to infect a *tet-s* cell which is lysogenic for P22, that is, *tet-s* (P22). In general, the *tet-r* allele is located at or adjacent to the site of the original P22 prophage.

When transductants *(tet-r* P22 lysogens) are induced by ultraviolet light, the cells lyse and release a normal number of phage-like particles which can be counted by using the electron microscope. These lysates have several interesting properties: (1) Only one is 10^5 particles can form a plaque; (2) when these phage-like particles infect a *tet-s* Salmonella, they transduce the *tet-r* allele at a frequency of 1.1 transductant per 10 particles; (3) when the lysate is used to infect *Salmonella* at various moi, the number of phage particles produced (counted by electron microscopy) depends strongly on moi, that is, at moi = 0.03, 0.3, 3; and 10, the numbers of phages produced per cell are 0.2, 20, 250, and 270, respectively. What is the nature of these phage-like particles? *Hint:* P22 packages DNA by the "headful" mechanism, that is, the length of DNA required to fill a phage head is cut from a concatenated phage DNA molecule.

15-75. Following are some facts about specialized transduction by phage P22 in *Salmonella typhimurium.*

1. Specialized transducing phages of P22 carrying the *proA* and *proB* genes can be generated. These genes are immediately adjacent to the prophage attachment site on the bacterial chromosome. The *proA-* and *proB*-specialized transducing phages have the behavior described below.

2. Lysates of the specialized transducing phages can go through the lytic cycle, but only in mixed infections.

3. In single infections the phages transduce by substitution; however, in mixed infections, the phages transduce by lysogenization.

4. In single infections of a particular recipient strain, unstable merodiploid transductants are observed in transductional crosses between complementing mutations in the proline operon of *S. typhimurium.*

5. In the presence of a functional recombination system, the transductants observed in (4) are unstable and segregate clones of both parental genotypes.

What is the molecular basis of each of the foregoing observations?

15-76. Two investigators meet each other at a scientific meeting and discover that they have been studying two similar phages: both phages are temperate and have the same appearance in the electron microscope. The phages differ in their host range (ability to adsorb to different *E. coli* strains); each carries a distinct immunity (repression) system; and each attaches to different sites on the *E. coli* chromosome. The properties of these two phages are compared in Table 15-4.

TABLE 15-4

	Phage Name	
Property	*186*	*P2*
Ability to adsorb to *E. coli* K	+	+
Ability to adsorb to *E. coli* C	−	+
Ability to form a plaque on *E. coli* K(186)	−	+
Ability to form a plaque on *E. coli* K(P2)	+	−
Location of prophage attachment site (determined by cotransduction)	Near *phe*	Near *ile*

(a) The investigators want to create a hybrid phage carrying the immunity specificity of phage 186 and the host range (adsorption properties) of phage P2. Design a selection for such a phage.

(b) Having obtained *h*P2 *imm*186 (the desired hybrid), the investigators wish to determine its chromosomal attachment site. How can this be done by bacterial mating?

(c) How can the chromosomal attachment site be located by transduction?

(d) In the experiment of part (c), 95 percent of the hybrid phage (*h*P2 *imm*186) tested attach near *phe*. What does this mean?

●**15-77.** *Salmonella* phage P22 is capable of generalized transduction. A transducing lysate is prepared on a *thi*+ *Salmonella* strain, then mutagenized with hydroxylamine and used to transduce a *Salmonella thi*− mutant at 30° C. Many of the *thi*+ transductant colonies are small and some of them cannot grow on replica plates at 42° C. When a similar transduction is performed with unmutagenized phage, the *thi*+ transductant colonies are more uniform in size, and all transductants grow at 42°. Can you explain the phenomenon?

15-78. Bacteriophage T5 particles labeled with radioactive DNA are adsorbed to *E. coli* in the presence of cyanide, which stops the generation of a cellular energy supply. After adsorption at 37°C for 20 minutes in the presence of cyanide, the infected cells are agitated in a blendor to remove the phage. The infected cells now contain radioactive DNA which can be extracted and which sediments as though its molecular weight were 10% that of a T5 DNA molecule.

Mutagenized T5 is adsorbed in the presence of cyanide and then removed by blending. The infected cells are superinfected with nonmutagenized T5 and allowed to lyse. Amber and temperature-sensitive mutants isolated from the burst represent only two of the twenty known T5 cistrons. From these data what can you say about (a) the mode of injection of T5 DNA and (b) the structure of T5 DNA?

15-79. Suppose you are studying a newly isolated temperate phage ϕ1971. You isolate nonsense mutations and classify them into four complementation groups: *a*, *b*, *c*, and *d*. You measure recombination after lytic growth of various pairs of mutant phage and obtain the data shown in Table 15-5.

TABLE 15-5

Parental phage	Percent wild-type recombinants
$a^- b^-$	3
$a^- c^-$	6
$a^- d^-$	1
$b^- c^-$	3
$b^- d^-$	2
$c^- d^-$	5

(a) From these data, draw the recombination map of the lytic phage.

(b) After growth of ϕ1975 on a new bacterial strain M6, you find that you have isolated a variant of ϕ1975 (termed ϕ1981) that will grow well on all hosts but will complement only b^- or c^- but not a^- or d^- mutants of ϕ1975. How do you think the new structure arose?

15-80. When phage T4 infects *E. coli*, phage-induced nucleases degrade the host DNA to mononucleotides. In a typical infection, about 300 phages are produced in each cell. Roughly how many phages would be produced by infection of a thymine-requiring *E. coli* by a T4 phage mutant lacking the thymidylate synthetase gene in a growth medium lacking thymine? Repeat for phages T7 and λ.

●**15-81.** In 1965, H. Ikeda and J. Tomizawa carried out the following experiments, which are explained in *J. Mol. Biol.* (1965), **14**:85–109. *E. coli* cells, which are genetically unable to synthesize thymine, were fed 5-bromouracil to make their DNA "heavy" and then were infected with the generalized transducing-phage P1*kc* and incubated until lysis in a medium containing (1) thymine, (2) 5-bromouracil, or (3) thymine plus radioactive phosphorus.

Analysis of phage progeny by density-gradient centrifugation showed that transducing particles from the first two media have similar densities and that those from the third are *not* radioactive.

From these observations, what do you conclude about the phage genetic material in transducing phages and the origin of transduced segments?

●15-82. Virulent phage species probably have higher recombination and mutation rates than temperate phage.

(a) What evolutionary pressures do you see for this situation?

(b) Propose three ways in which virulent phage might achieve a higher mutation rate (be as explicit and biochemically reasonable as you can and choose mechanisms as diverse as you can).

●15-83. Bacteriophage KR is a temperate virus whose host is *Exemplis gratis* (a bacterial species quite like *E. coli*). Mutants of KR are known which affect the immunity repressor of the phage and are therefore totally unable to lysogenize. These mutants have been mapped in the *tul* (*t*otally *u*nable to *l*ysogenize) gene.

In addition, mutants of KR can be isolated which lysogenize rarely. These mutants map in the *dilA* and *dilB* genes (*d*efective *i*n *l*ysogeny). (Furthermore, it is known from other experiments that *dil* mutants are not defective in the enzymes which catalyze the insertion of KR into the *E. gratis* chromosome). The data for frequency of lysogeny for each of these mutants in *E. gratis* are shown in Table 15-6. Based on the foregoing data, provide answers for the following:

TABLE 15-6

Infecting phage or phages	Relative percent of lysogenic survivors	Genotype of prophage
tul⁺ *dilA*⁺ *dilB*⁺	100	*tul*⁺ *dilA*⁺ *dilB*⁺
tul⁻	0	—
dilA⁻	0.1	*dilA*⁻
dilB⁻	1.0	*dilB*⁻
tul⁻ + *dilA*⁻	60	*dilA*⁻ or *tul*⁻ *dilA*⁻ dilysogen
tul⁻ + *dilB*⁻	75	*dilB*⁻ or *tul*⁻ *dilB*⁻ dilysogen
dilA⁻ + *dilB*⁻	80	*dilA*⁻ or *dilB*⁻ or *dilA*⁻ *dilB*⁻ dilysogen

(a) What process is going on when mixed infections of *tul⁻* and *dilA⁻* or *dilB⁻* yield lysogens?

(b) Provide a reasonable model for the role of the *dil* genes in lysogenization.

(c) Why are single lysogens of *tul⁻* not isolated after mixed infection, while single *dilA⁻* or *dilB⁻* lysogens can be found?

●15-84. Generalized and specialized transducing phages carrying an amber suppressor tRNA gene are used to transduce a recipient strain carrying an amber mutation in the β-galactosidase gene. Assume that the recipient has only one copy of the wild-type (non-suppressing) tRNA gene, that it is allelic with the amber suppressor tRNA gene of the donor, and that the transductants which are being scored are only those that result from suppression of the β-galactosidase amber mutation. Why is the generalized transducing phage preparation unable to transduce the recipient but the specialized transducing phages are able to do so? Would the generalized transducing phages be able to transduce the recipient abortively?

15-85. Could either RNA or single-stranded DNA phages carry out specialized transduction?

●15-86. For this problem, read the article by J. Ebel-Tsipis, D. Botstein, and M. S. Fox, *J. Mol. Biol* (1972), **71**: 433–448. How did they prove, by genetic experiments, that the donor DNA and recipient DNA molecules are most likely covalently linked in generalized transduction?

15-87. Why are P22 specialized-transducing particles generated only by induction of donor cells rather than by lytic infection?

●15-88. For *E. coli* phage φX174, would you expect the + and − circular, single-stranded DNA strands of phage φX174 (isolated from the double-stranded replicative form) each to be infective in transfection experiments? Would you also expect the + and − RNA strands of the *E. coli* RNA phages to be individually infective? State your reasons.

●15-89. For this problem, see P. D. Baas and H. S. Jansz, *J. Mol. Biol.* (1972), **63**: 569–576. In the experiments by Baas and Jansz dealing with the origin of replication of the RF (replicative form) of φX174, the wild-type marker was always present on the − strand and the mutant marker was always on the + strand of the DNA heteroduplex. Why were the heteroduplexes used in the transfection experiments prepared in this way?

●15-90. For this problem, see *Proc. Nat. Acad. Sciences, U.S.A.* (1975), **72**: 235. What is the evidence that figure-eight DNA structures are intermediates in recombination of φX174? Using density transfer experiments and appropriate host and phage mutants, how could you attempt to prove whether or not circular dimers of φX174 DNA are produced by recombination and/or by replication?

15-91. How would you determine which cistrons are encompassed in ϕX174 RF fragments generated by restriction enzymes?

15-92. What do you consider to be the strongest evidence (from studies with the single-stranded, circuuar DNA molecules from phages) for the rolling circle model of DNA replication?

15-93. A new temperature-sensitive *E. coli* mutant, *dnaQ*, affects host DNA replication at an elevated, nonpermissive temperature. How would you show whether the product of the *dnaQ* gene is necessary for any of the three stages of DNA replication of ϕX174?

15-94. List at least five differences between the infectious cycles *in vivo* of phages ϕX174 and the filamentous phages such as M13.

15-95. What do you predict the results would be if one carried out a Hershey-Chase experiment with labeled M13 phage?

●15-96. For phages containing single-stranded DNA, for example, ϕX174, would you expect genetic recombination to occur before conversion to their RF forms in multiply infected cells? Would you expect the RNA phages to recombine at the single-stranded or the RF stages of the infectious cycle?

●15-97. Which phage would be expected to produce clearer plaques, ϕX174 or M13? Why?

16
Microbial Genetics

Introduction

For decades the study of biology proceeded by observing things as they are. A major conceptual change occurred in the 1940's when it was realized that a great deal of significant information can be obtained by studying mutants. This was a major breakthrough in the elucidation of metabolic pathways. An important modification of this methodology was made by the microbial genetics group at the Pasteur Institute in Paris, in a study of regulation of the enzymes involved in lactose utilization in *E. coli;* the group introduced the idea of *strain construction* by way of *genetic recombination*—that is, the genotype of a bacterium could be deliberately modified to obtain specific information.

Microbial genetics is of course interesting in its own right and its study is providing important information about the mechanism of genetic recombination.

There are numerous excellent texts which describe microbial genetics in detail. Thus, this introduction will consist of brief outlines of the processes, a few definitions, and some examples of strain construction.

PHAGE GENETICS

When two or more phages simultaneously infect a bacterium, genetic recombination is a frequent event. Under certain circumstances, owing to some of the peculiarities of phage maturation, recombination is obligatory. If looked at in the most general way, phage recombination appears to obey the rules of classical genetics described in Chapter 1—that is, the probability of recombination is proportional to distance between genetic markers and in a cross between a *population* of a^+b^- and a^-b^+ phages, the number of a^+b^+ recombinants equals the number of a^-b^- recombinants. When examined more closely, phage recombination exhibits the following complexities.

(1) Some phages, for example, *E. coli* phage λ, have two recombination systems and may also use the bacterial system (Rec). One of the λ systems utilizes the Red genes; this catalyzes recombination between any pair of genetic markers, yet because of asymmetries inherent in the phage packaging system and because all DNA is not packaged, recombinants are recovered more efficiently from some regions of the DNA than from others. The other λ system (Int) utilizes the enzyme, integrase, which catalyzes the site-specific exchange between the attachment sites *(att)* in the phage and bacteria involved in prophage insertion. The Int system can also make exchanges between two phage attachment sites and hence influences recombination if two genetic markers are on opposite sides of *att*.

(2) Whereas, if averaged over a population, the number of a^+b^+ and a^-b^- recombinants in a cross $a^+b^- \times a^-b^+$ are equal, this is rarely so when the phage progeny from one infected bacterium are examined and frequently one of the two types is even missing.*

(3) In crosses between parents of genotypes $a^+b^-c^+$ and $a^-b^+c^-$ where the gene order is *abc*, and the markers are well separated, the frequency of production of $a^+b^+c^+$ is that expected from the genetic distances *ab* and *bc*. If the markers are very close, the frequency of $a^+b^+c^+$ is very much greater than expected (that is, the number of double crossovers is very high). This is called *high negative interference.*

(4) If a population of bacteria is mixedly infected with two genetically marked phage types, so that on the average each bacterium is multiply infected, the multiplicity of infection of each bacterium by

---

* The following terminology is used to describe this phenomenon. The recombinants a^+b^- and a^-b^+ are called *reciprocal recombinants.* When these recombinants are recovered in equal numbers, it is said that recombination is reciprocal; otherwise, it is nonreciprocal. Thus, recombination may be reciprocal in a *mass lysate* yet nonreciprocal in a *single burst.*

each phage type follows a Poisson distribution—that is, all infected cells are not the same.

Phages have been very valuable in the study of genetic recombination. For example, mutations can be introduced into both phage and bacterium to alter recombination efficiency, DNA synthesis, and maturation. Furthermore, the use of multiple markers distributed along the genetic maps allows one to look closely at selected regions as well as at the entire chromosome at one time. For instance, A. H. Doermann has performed crosses with *E. coli* phage T4 in which 30 genetic markers were present in a single phage; 8 markers were clustered in a small region and 22 were distributed uniformly throughout the genetic map. Finally, phages can be examined by the elegant physical techniques of molecular biology, as is being done by Franklin Stahl and his associates, who perform crosses between density-labeled particles.

PHAGE CONSTRUCTION—AN EXAMPLE

Strain construction requires a battery of tricks used to enable selection of particular markers. For instance, let us construct $\lambda xis^-c^+chi^-$ from $\lambda red^-c^-chi^-$ and $\lambda xis^-c^+S^-$:

xis^+ red$^-$	c^-		chi^-S^+
xis^- red$^+$	c^+		chi^+S^-

The *xis* and *red* genes are very near, as are *chi* and *S*; λc^+ and λc^- have turbid and clear plaques, respectively. A λred^- mutant fails to form a plaque on a ligase-deficient *(lig$^-$)* host; S^- fails to plate on a strain lacking a suppressor *(sup$^-$)*. Neither *xis$^-$* nor *chi$^-$* have easily recognized phenotypes.

The two phages are crossed and the lysate is plated on a *lig$^-$sup$^-$* bacterium. Only the phages which are *red$^+$S$^+$* can form a plaque. Some of these plaques will be clear and some turbid depending upon whether the exchange is to the left or right of the c^- allele. Since the *xis* and *red* genes are close, the *red$^+$* phages are probably *xis$^-$*; similarly, the S^+ phages are probably *chi$^-$*. Thus the clear plaques are likely to be $xis^-c^-chi^-$. Further tests for the *xis$^-$* and *chi$^-$* characters should be carried out; these might be tedious, but because only one or two plaques need to be tested, this is not a hard task.

BACTERIAL GENETICS

Genetic manipulations in bacteria are accomplished either by mating or by transduction, mating being the more general method.

Some strains of *E. coli* contain a circular DNA molecule known as the ***sex factor,*** fertility factor, or simply F. This is a transmissible element carrying information for both replication of its DNA and the ability to transfer a replica of itself to a recipient cell. F can become stably integrated in the host DNA; a cell containing F in this state is called an ***Hfr.*** Analogous to specialized transducing phages (see Chapter 15), F can sometimes be excised and carry with it host DNA from the region immediately adjacent to its insertion site. An element carrying chromosomal DNA is called an ***F' factor;*** if it carries genes x, y, and z, it is written F'xyz.

There are many chromosomal sites of insertion of F and hence many classes of Hfr cells. Each Hfr is capable of transferring its chromosome to a female cell by starting at a site adjacent to F. The direction of transfer is such that F is transferred last, if at all. When cultures of Hfr and female cells are mixed, Hfr : female pairs form and DNA transfer begins. Owing to the disrupting effect of Brownian motion and other physical disturbances, the mating couple usually breaks apart before the entire chromosome is transferred. Sometimes the culture is shaken violently to separate the pairs—this is called ***interrupted mating.***

In a typical mating, an Hfr whose genotype is $a^+b^+c^+$ (and so on) transfers DNA to a female whose genotype is $a^-b^-c^-$. If the cells are placed on agar containing nutrients B and C, only cells which are a^+ can form colonies; a^+ is called the ***selected marker.*** Two types of a^+ cells will grow, the Hfr and a female which has received the a^+ allele from the Hfr. To prevent the Hfr from growing, a second marker is added. The most common marker is streptomycin, with the Hfr and the female chosen to be streptomycin-sensitive *(str-s)* and streptomycin-resistant *(str-r)*, respectively. Thus, if streptomycin is included in the agar, no Hfr cells can grow; a female recombinant will grow as long as it has not received the *str-s* allele from the Hfr. The marker used to select *against* the Hfr is called the ***counter-selected marker.***

Consider a mating between the Hfr with genotype a^+b^-str-s and a female which is a^-b^+str-r. We may select a^+str-r recombinants and ask how often they are also b^+. When a marker b is examined in this way, it is called an ***unselected marker.***

CONSTRUCTION OF A BACTERIAL STRAIN—AN EXAMPLE

We have a female strain which is *lac⁻str-r* and also carries a mutation in *gal* called *gal*1; we need a strain which is *lac⁻gal*2. We have two *str-s* Hfr strains which transfer *gal* before *lac*—one is *gal⁺lac⁺*, the other, *gal* 2 *lac⁺*. Color-indicator agar is available in which *gal⁺* colonies are white and *gal⁻* colonies are red. We cannot use the cross Hfr *gal* 2 *lac⁺* × *gal* 1*lac⁻*, for there is no way to distinguish *gal* 1 and *gal* 2 in agar. Therefore, we cross Hfr *gal⁺lac⁺str-s* with the *gal* 1 *lac⁻str-r* female, interrupt the mating before the *lac⁺* allele is transferred, and put the cells on Gal-indicator agar containing streptomycin. We select several white colonies—these are *gal⁺*—and test each to see that it still is *lac⁻*. This new *gal⁺lac⁻str-r* female is taken and mated with the *gal* 2 *lac⁺str-s* Hfr. Again a short mating is done; a red colony is selected and tested to see if it is *lac⁻*. It will be a *gal* 2 *lac⁻str-r* recombinant.

STRAIN CONSTRUCTION BY WAY OF TRANSDUCTION

If one has a *leu⁻lac⁻* cell and wishes to convert it to *leu⁺lac⁻*, it is a simple matter to grow phage P1 on a *leu⁺* host, infect the *leu⁻lac⁻* cells with the P1 lysate, put the infected cells on agar lacking leucine, and select a colony which grows. It is slightly more complicated to alter a particular strain so that it is defective for polymerase I, that is, *polA⁻*, because there is no positive selection for the *polA⁻* allele. To do this one uses **cotransduction** of an unselected marker. For instance, the gene *metE* is so near *polA* that they are frequently cotransduced. To start, the desired strain should be made *metE⁻*; this can be done by mutation, or by Hfr mating in which *metE⁻* is used as an unselected marker. Phage P1 is then grown on any strain which is *metE⁺polA⁻* and the lysate is used to transduce the *metE⁻* strain to *metE⁺*. Colonies which grow on agar lacking methionine, that is, the *metE⁺* colonies, are tested for the *polA⁻* allele by transferring a part of them to agar containing methyl methane sulfonate. Those which fail to grow are *polA⁻*.

REFERENCES

Davis, B. D., R. Dulbecco, H. N. Eisen, H. S. Ginsberg, and W. B. Wood. 1973. *Microbiology*. New York. Harper & Row. Chapters 9 and 45.

Hayes, W. 1968. *The Genetics of Bacteria and Their Viruses*. New York. John Wiley & Sons.

Herskowitz, I. H. 1973. *Principles of Genetics*. New York. Macmillan. Chapters 10, 11, 12, and 13.

Jacob, F., and E. L. Wollman. 1961. *Sexuality and the Genetics of Bacteria*. New York. Academic Press. The first book on the subject, and still worth reading.

Stahl, F. W. 1964. *The Mechanics of Inheritance*. Englewood Cliffs, N. J. Prentice-Hall.

Stent, G., and R. Calendar, 1978. *Molecular Genetics*. San Francisco. W. H. Freeman and Company. Chapters 9, 11, and 12.

Suzuki, D. T., and A. J. F. Griffiths. 1976. *An Introduction to Genetic Analysis*. San Francisco. W. H. Freeman and Company. Chapter 6.

Problems

○**16-1.** Approximately 10^8 *E. coli* cells of a mutant strain are plated on complete medium, and form a bacterial lawn. Replica plates are prepared containing minimal medium supplemented by leucine, thymine, and proline, as shown below.

(a) What is the genotype of the mutant strain?

(b) Explain the colonies which appear on the replica plates.

Complete medium   Minimal medium supplemented by

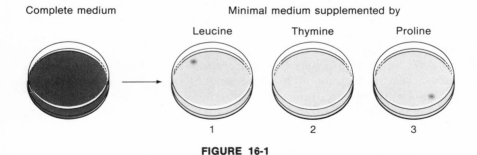

FIGURE 16-1

○**16-2.** An Hfr cell with the genotype $arg^+leu^+pro^+thr^+thy^-ser^-his^-str\text{-}s$ is mated with an F^- cell which has the genotype $arg^-leu^-pro^-thy^-ser^-his^-str\text{-}r$. Recombinants are selected by plating on agar containing all seven nutrients plus streptomycin. This master plate is replica-plated onto the six plates shown in Figure 16-2, with the indicated supplements. Colonies grew as shown. What are the genotypes of the numbered colonies?

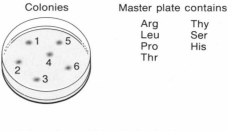

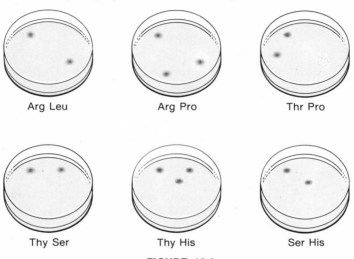

FIGURE 16-2

○**16-3.** Bacteriophage T4 has a gene that controls the structure of an enzyme called lysozyme, which initiates lysis of the infected bacterial cell. This enzyme molecule consists of only one polypeptide chain. Would any mutations in this gene be expected to show *intragenic* complementation? Explain.

○**16-4.** Phage P1 is grown on a bacterium with genotype $a^+b^+c^+$. The resulting P1 is used to infect a bacterium which is $a^-b^-c^-$, and various transductants are examined. Of those which are b^+, 80 percent are also a^+ but only 1 percent are also c^+. Of those which are c^+, the number which is also a^+ is greater than the number which is b^+. What is the gene order?

○**16-5.** Using P22 as a generalized transducing phage grown on a $pur^+pro^-his^-$ donor, a recipient strain having genotype $pur^-\,pro^+\,his^+$ was infected and pur^+ transductants were selected.
(a) What nutrients were in the plate used for selection?
(b) The pur^+ transductants were tested for the unselected *pro* and *his* markers, and the data in Table 16-1 were found.

TABLE 16-1

Genotype	Number of colonies
$pro^+\,his^+$	103
$pro^-\,his^+$	24
$pro^+\,his^-$	158
$pro^-\,his^-$	1

What is the gene order and what is the relative distance between the mutations?

16-6. Lac^+ colonies are purple on EMB agar; lac^- are pink. A strain, S, is found which gives 99 percent purple colonies and 1 percent pink, when grown at 30° C. If S is grown at 42° C for several generations and then plated at 30°, half of the colonies are purple and half are pink. When pink colonies from a 42° plate are restreaked at 30°, only pink colonies result; when purple colonies from a 30° plate are replated, the colonies are purple at 30° and pink at 42°. At a frequency of 1 in 10^4 a purple colony can be found on a plate maintained at 42°. Such purple colonies fall into three classes: (1) Those which remain purple at 42° but still produce about 1 percent pink colonies at 30° or 42°; (2) Those which are purple at 42° but never (<1 in 10^5) produce pinks at 42° or 30°; (3) Those which are like class (2) but have the additional property of being resistant to phage T1 (the original strain was T1-sensitive). Other classes are possible but will not be considered here. What is the relevant genotype of the original strain S? What are the genotypes of the three classes of variants? *Note:* T1-sensitivity is dominant over T1-resistance.

●16-7. How would you go about trying to prove whether streptomycin-resistance (acting at the ribosomal level) is dominant or recessive? You are only given a streptomycin sensitive F^+ strain, and appropriate media and materials. Write out a protocol, including appropriate controls.

○16-8. An Hfr $a^+b^+c^+d^+str\text{-}s$ is mated with an F^- strain that is $a^-b^-c^-d^-str\text{-}r$. Mating is interrupted at various times and each sample that is obtained is plated on four different agar types on which $a^+str\text{-}r$, $b^+str\text{-}r$, $c^+str\text{-}r$, and $d^+str\text{-}r$ recombinants can grow. A plot of the number of recombinants of each type as a function of time generates a set of *time-of-entry curves,* so called because the intersection of each line with the time axis defines the time at which a particular allele first enters an F^- cell.

Consider the time-of-entry curves shown in Figure 16-3.

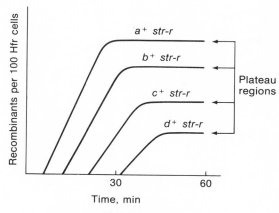

FIGURE 16-3

(a) Which of the following statements (1 or 2) is the explanation for the fact that the curves form *plateau regions* at lower values as one proceeds rightward on the graph?

(1) The probability of a gene being inserted into the F^- chromosome decreases with time.

(2) The probability of a gene being transferred to the F^- cell decreases with time.

(b) If the Hfr is a λ lysogen, it is often found that after a certain time of entry (that of the λ prophage), there are no recombinants. This is because the λ prophage is induced when entering the female since the female contains no repressor. Suppose λ is transferred starting at 15 minutes. Will there be a lowering (with respect to the y-axis) of the plateau region of a gene which is normally transferred starting at 5 minutes?

(c) If the female is also a lysogen, will the time-of-entry curves be the same as would occur if neither Hfr nor female are lysogens? State the reason for your answer.

16-9. In a cross between an Hfr that has the genotype $leu^+lac^+gal^+trp^+$ str-s and a female having genotype $leu^-lac^-gal^-trp^-str$-r, the time-of-entry curves in Figure 16-4 were obtained. Draw what you would expect to happen if

(a) the Hfr was a λ lysogen and the female is not.

(b) both the Hfr and the female are λ lysogens.

Note: Remember that the λ prophage is located between *gal* and *bio*. Briefly explain your answer.

16-10. Consult the genetic map of *E. coli* on page 241. Suppose an F factor were inserted into the *strA* gene of *E. coli*. If the resulting Hfr strain was then shown to transfer the *fda* gene within a few

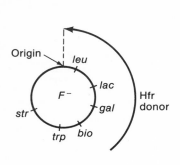

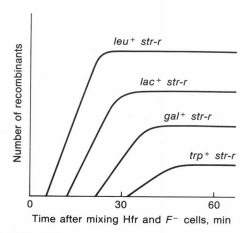

FIGURE 16-4

minutes after the onset of conjugation, at what time after conjugation started would this strain transfer the *argC* gene?

16-11. In a mixture of Hfr P4 donors with genotype *pro⁺mal⁺lac⁺* and cell recipients with genotype *F⁻ pro⁻ mal⁻ met⁻ lac⁻*, a few recombinants arise that carry only the *pro⁺* marker. All of these recombinants are Hfr donors. Account for this result and state the location of F in Hfr P4.

○16-12. Explain why some Hfr strains transfer their genes to an *F⁻* in a clockwise order whereas some do so in a counterclockwise order.

16-13. When an *F⁺* cell mates with an *F⁻* cell, the recipient is converted to a male with very high efficiency. This is not so when the donor is an Hfr. How do you explain this?

●16-14. Assume the *E. coli* map is in alphabetical order, *a–z*. An Hfr known to transfer in the order *a* to *z*, and which is wild-type (+) for all genes, is mated to a female which is *a⁻m⁻z⁻* (all genes are equally spaced, that is, the distance *a–b* = *j–k* = *r–s* = *y–z*, and so on). The Hfr is streptomycin-sensitive (*str-s*); the female is *str-r*. The *str* locus is far from *a, m,* and *z*. Mating is allowed to proceed for a long time. When *m⁺str-r* recombinants are selected, about ¼ are also *a⁺* and about 0.02 are also *z⁺*. Most *a⁺str-r* cells are *m⁻* and *z⁻*. However, most *z⁺str-r* recombinants are also *a⁺*. How might you explain this finding?

16-15. Suppose you have isolated two *independent* arginine-requiring (*arg⁻*) mutant strains from a parent *E. coli* strain which already required methionine (*met⁻*) and was resistant to streptomycin (*str-r*). You mate your two mutants (1 and 2) with an Hfr strain which is streptomycin-sensitive (*str-s*) and does not require arginine or methionine for growth (that is, *arg⁺met⁺*). Using the interrupted mating technique you obtain the time-of-entry curves shown in Figure 16-5. Draw a genetic map which explains the difference observed in the two matings. (See problem 16-8 for a description of time-of-entry curves.)

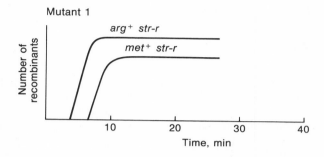

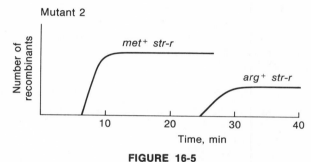

FIGURE 16-5

16-16. You are interested in the biosynthetic pathway of compound X and isolate 10 different (independently isolated) mutants of *E. coli* which require compound X for growth. The mutations are mapped by conjugation and their approximate positions are given in Figure 16-6.
(a) What is the minimum number of genes involved in the synthesis of compound X?
(b) Why must your answer be the minimum estimate?

○**16-17.** The gene order in an Hfr strain is *abcd*. In a cross between an Hfr donor that has genotype $a^+b^+c^+d^+x^-str$-s and a female that is $a^-b^-c^-d^-x^+str$-r, 90 percent of the d^+str-r recombinants are x^- and 100 percent of the c^+d^+str-r recombinants are x^-. The times of entry of *a*, *b*, *c*, and *d* are 5, 10, 15, and 20 minutes; *str* enters at 55 minutes. Where is *x* located?

○**16-18.** Consider an Hfr donor that has genotype $a^+b^+c^+str$-s mating with a female that has genotype $a^-b^-c^-str$-r; the order of transfer is

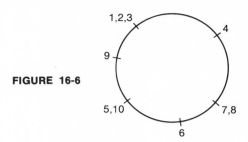

FIGURE 16-6

abc. Suppose *a*, *b* and *c* are far enough along that one does not have to worry about *a* being transferred very early. Also the distance between *a* and *b* is the same as the distance between *b* and *c*. None of the markers are near *str*. Recombinants are selected as usual by plating on agar lacking X and containing streptomycin. Which of the following are true (several answers are)? Explain.

(a) a^+str-r colonies $> c^+str$-r colonies.

(b) b^+str-r colonies $< c^+str$-r colonies.

(c) a^+b^+str-r colonies $< b^+str$-r colonies.

(d) a^+b^+str-r colonies $= b^+str$-r colonies.

(e) Most a^+c^+str-r colonies will also be b^+.

(f) Most b^-c^+str-r colonies will also be a^-.

(g) $a^+b^+c^-str$-r colonies $< a^+b^-c^-str$-r colonies.

16-19. Which of the following crosses would you carry out to determine the location of prophage λ, and why?

(a) Hfr (λ) × F⁻;

(b) Hfr (λ) × F⁻ (λ);

(c) Hfr × F⁻;

(d) Hfr × F⁻ (λ).

16-20. An Hfr with genotype $a^+b^+c^+d^-str$-s is mated with a female with genotype $a^-b^-c^-d^+str$-r. At various times the culture is shaken violently to break apart pairs and the cells are plated on agar having the composition shown in Table 16-2 (+ means that the nutrient is present). The numbers of colonies growing on the agar are shown in Table 16-3. One hundred colonies from the 25-min plates are picked and transferred to a dish containing agar of type 4, which contains A, B, C, streptomycin, but no D. The number of colonies which grow on type 4 is shown in Table 16-4.

(a) What is the order of the genes *a*, *b*, *c*, and *d*?

(b) Roughly how many colonies would be expected at various times on agar containing C and streptomycin but not A or B?

TABLE 16-2

Agar of type	Str	A	B	C	D
1	+	+	+	−	+
2	+	−	+	+	+
3	+	+	−	+	+

TABLE 16-3

Time of sampling, min	Number of colonies on agar of type		
	1	2	3
0	0	0	0
5	0	0	0
7.5	100	0	0
10	200	0	0
12.5	300	0	75
15	400	0	150
17.5	400	50	225
20	400	100	250
25	400	100	250

TABLE 16-4

Colonies taken from agar of type	Number of colonies on agar of type 4
1	89
2	51
3	8

○16-21. Suppose you are given a mutant strain of *E. coli*. In checking it out you test growth requirements at two temperatures. You plate 10^8 cells at either 25° C or 42° C on minimal agar plates supplemented with various amino acids. The amount of growth is indicated in Table 16-5.

TABLE 16-5

Supplement	Temperature (°C)	Growth of colonies
His, Trp	25	None
His, Leu	25	10
Leu, Trp	25	Confluent
His, Leu, Trp	25	Confluent
His, Trp	42	None
His, Leu	42	8
Leu, Trp	42	12
His, Leu, Trp	42	Confluent

(a) What is the genotype of the strain?

(b) Why is there confluent growth on the plate containing leucine and tryptophan at 25° C but not at 42° C?

(c) Why are there *no* colonies on the plates containing histidine and tryptophan?

(d) What is the expected genotype of the colonies that grow at 42° on the agar containing histidine and leucine?

(e) What is the expected genotype of the colonies that grow at 42° on the agar containing leucine and tryptophan?

(f) Suppose you have an Hfr strain with genotype $met^-his^+leu^+$ trp^+. This Hfr is known to transfer *met* very late. You mix 10^8 bacteria of the Hfr strain and 10^9 of your mutant bacteria and allow mating to occur for several hours. You plate a 10,000-fold dilution of the mixture on the following plates at 42° C and incubate them for two days. You observe the results shown in Table 16-6.

TABLE 16-6

Nutrient or nutrients in agar	Number of colonies found
Trp, His	250
Leu, His	50
Leu, Trp	500
His	10

(1) Which genes entered first, second, and third?

(2) You now know the *relative* order of these 3 markers, but you know nothing about their exact *location* on the chromosome. Describe an experiment (other than transduction) that will tell you where these markers are located on the chromosome.

(3) What is the purpose of the *met⁻* mutation in the Hfr strain?

●**16-22.** Suppose you collect a large number of galactose-negative (*gal⁻*) bacterial mutants and identify three closely-linked genes (designated *galA, galB, galC*) by complementation and rough mapping studies. You wish to order these genes and learn something about the genetic structure of the "galactose region" of DNA.

(a) To order the genes, you use Hfr × F⁻ mating between an Hfr with genotype *bio⁺gal⁺str-s* and various F⁻ strains with genotype *gal⁻bio⁻str-r*. The Hfr transfers *bio* after *gal*. For the mapping, you select *bio⁺str-r* recombinants and score the frequency of *gal⁺* among the *bio⁺str-r* recombinants. You obtain the data shown in Table 16-7. What is the gene order relative to *bio*?

TABLE 16-7

gal mutation	Number of gal⁺/number of bio⁺ in cross
galA⁻	0.65
galB⁻	0.72
galC⁻	0.84

(b) Among your collection of *gal⁻* mutations, you have heat-sensitive (*gal⁻* only at high temperature), cold-sensitive (*gal⁻* only at low temperature), and nonsense (chain-termination) mutants. The distribution of these mutants among genes is as shown in Table 16-8.

TABLE 16-8

Class of mutation	Distribution		
	galA	galB	galC
Nonsense	20	50	100
Heat-sensitive	5	100	5
Cold-sensitive	1	0	20

From these data, estimate the relative size of the three genes (assume the mutation rate per base pair is constant). Justify your answer briefly.

16-23. Suppose you are mating an Hfr donor of genotype $a^+b^+c^+d^+str$-s with an F^- recipient having genotype $a^-b^-c^-d^-str$-r. Genes a, b, c, and d are spaced equally. A time-of-entry experiment is carried out and the data shown in Table 16-9 are obtained.

TABLE 16-9

Time of mating, in min	Number of recombinants of indicated genotype per 1,000 Hfr			
	a^+str-r	b^+str-r	c^+str-r	d^+str-r
0	0.01	0.006	0.008	0.0001
10	5	0.1	0.01	0.0004
15	50	3	0.1	0.001
20	100	35	2	0.001
25	105	80	20	0.1
30	110	82	43	0.2
40	105	80	40	0.3
50	105	80	40	0.4
60	105	81	42	0.4
70	103	80	41	0.4

What are the times of entry for each gene? Explain the low recombination frequency in the plateau region for d^+str-r recombinants.

16-24. (a) A bacterial culture was infected with an input of 3 T2h^- (host-range) and 17 T2m^- (minute-plaque-forming) phages per cell. After the culture lysed, the progeny were plated and scored, and 7 percent were found to be wild-type. What frequency of h^-m^- recombinants were found?
(b) The cross was repeated using an input of 10 T2h^- and 10 T2m^-. Were more, or fewer, wild-type found than were found in part (a)?
(c) What frequency of wild-type were found in (b)?

16-25. In early work with phage λ (before the discovery of the integration system), 12 percent recombination was observed in equal-input high-multiplicity crosses between the terminal markers $m6$ and mi^-. Certain features of the cross suggested that in individual matings, $m6$ and mi^- were in fact assorting essentially independently (in the Mendelian sense). With these considerations, the average number of matings in the ancestry of a λ progeny phage particle can be estimated by using the Visconti-Delbrück theory. What is this value?

●**16-26.** Suppose you are studying a newly isolated temperate phage $\phi 177$. The phage can integrate into the bacterium DNA very near the genes for lactose utilization *(lac)*. You isolate nonsense mutations and classify them into four complementation groups a, b, c, and d.
(a) You measure recombination, after lytic growth of various pairs of mutant phages, and obtain the data shown in Table 16-10. From these data, draw the recombination map of the vegetative (lytic) phage.

TABLE 16-10

Parental phage	Percent wild-type recombinants
a^-b^-	12
a^-c^-	6
a^-d^-	1
b^-c^-	6
b^-d^-	11
c^-d^-	5

(b) For the prophage, you measure P1 cotransduction of *lac*⁺ and the wild-type alleles of a^-, b^-, c^-, or d^- (for example, P1 grown on $lac^+a^+b^+c^+d^+$ is used to transduce $lac^-a^-b^+c^+d^+$ and the frequency of $lac^+a^+b^+c^+d^+$ is measured). The results are shown in Table 16-11. From these data, draw the prophage map.
(c) What is the most likely physical structure of the DNA isolated from the phage (for example linear, or circular, and what nature of ends, if linear). Propose a physical experiment with isolated DNA to test your supposition.

TABLE 16-11

Initial prophage genotype	Percent wild-type among *lac*⁺ transductants
$a^-b^+c^+d^+$	20
$a^+b^-c^+d^+$	30
$a^+b^+c^-d^+$	40
$a^+b^+c^+d^-$	16

●**16-27.** Suppose that phage mutants are isolated which show decreased ability to recombine. All the mutants fall into two complementation groups, *recX* and *recY*. The phage has the following map order: $defghjklmn$. Each of the 10 genes is one recombination unit apart (that is, $d \times e = 1$ percent, $j \times k = 1$ percent, and so on). Crosses are now performed with rec^- mutants. The data in Table 16-12 are obtained.

TABLE 16-12

Genotypes of parental phages in crosses	Genotype of recombinant	Recombination frequency, percent
$d^-\,recX^-,\,e^-\,recX^-$	$d^+\,e^+$	0.0001
$f^-\,recX^-,\,j^-\,recX^-$	$f^+\,j^+$	0.0001
$j^-\,recX^-,\,m^-\,recX^-$	$j^+\,m^+$	1
$j^-\,recX^-,\,n^-\,recX^-$	$j^+\,n^+$	1
$k^-\,recX^-,\,m^-\,recX^-$	$k^+\,m^+$	1
$k^-\,recX^-,\,l^-\,recX^-$	$k^+\,l^+$	0.5
$j^-\,recX^-,\,k^-\,recX^-$	$j^+\,k^+$	0.0001
$l^-\,recX^-,\,m^-\,recX^-$	$l^+\,m^+$	0.0001

If the mutants are $recY^-$, all recombinants occur with >1 percent frequency. What properties of the $recX$ and $recY$ systems are suggested by these results?

16-28. Consider a T4 amber mutant, $am\,82$, which plates on the *E. coli* sup^+ strain CR63 but not on *E. coli* strain B sup^-, whereas T4 am^+ plates on both. It is possible to prepare a lysate of T4 am^+ having the following strange plating properties.
(a) If 10^8 T4 am^+ are adsorbed to 10^9 *E. coli* B, and plated on strain CR63, the phage titer is 1.0×10^8/ml.
(b) If 3×10^9 T4 $am\,82$ are adsorbed to 10^9 *E. coli* B and plated on strain CR63, the phage titer is 3×10^7/ml. Why is this so low?
(c) If 10^8 T4am^+ *and* 3×10^9 T4$am\,82$ are simultaneously adsorbed to strain B and then plated on strain CR63, the phage titer is 3.5×10^8/ml. What is the significance of this observation? In other words, what special type of phage must the T4 am^+ lysate contain?

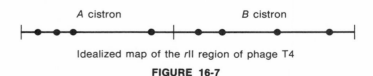

Idealized map of the *rII* region of phage T4

FIGURE 16-7

16-29. Assume that Figure 16-7 is an idealized map of the *rII* region of phage T4. Several point mutations are indicated by solid circles. A new *rII* mutation is found *(z)* that does not complement any of the above mutations, and it does not revert. In genetic crosses, *z* recombines with any of the above mutations and produces wild-type phages. What explanation can you suggest?

16-30. A temperate phage is treated with the mutagen nitrosoguanidine and five clear-plaque mutants are picked for further study. To study the lysogenic pathway of phage development, the clear mutants are coinfected pairwise with each other and with the turbid wild-type phage c^+. The frequency of lysogeny is measured, with the results shown in Table 16-13.

TABLE 16-13

	c_1	c_2	c_3	c_4	c_5	c^+
			Percent lysogeny			
c_1	2	59	57	1	50	52
c_2		1	0	56	52	60
c_3			1	57	58	55
c_4				1	56	57
c_5					0	55
c_6						60

(a) What genetic method does the data demonstrate?

(b) How many cistrons are there? Which mutations are in the same cistrons?

(c) Can a genetic map be derived from the data given in Table 16-13?

(d) Phage mutant c_5 is known never to lysogenize its host. What is a possible function for the wild-type product of the gene defined by the c_5 mutation (for example, is its function to establish or to maintain lysogeny)?

(e) Why do the mutants c_1 through c_5 form clear plaques, whereas c^+ forms a turbid plaque?

(f) If a lysogen of this wild-type phage is infected with an unrelated temperate phage, what will be the outcome?

●16-31. One amber mutant and six *ts* mutants of phage X have been isolated. Complementation tests result in the burst sizes shown in Table 16-14 (given as phages per cell after infection at 42° C in a sup^- host).

(a) Which mutant is most leaky? ("Leaky" is defined on p. 86.)

(b) How many cistrons are defined?

(c) Two-factor crosses give the recombination frequencies shown in Table 16-15 (in percent ts^+am^+). Draw a genetic map of phage X.

TABLE 16-14

	$am\,1$	$ts\,1$	$ts\,2$	$ts\,3$	$ts\,4$	$ts\,5$	$ts\,6$
$am\,1$	0.01						
$ts\,1$	100	1					
$ts\,2$	150	50	0.05				
$ts\,3$	20	75	10	0.1			
$ts\,4$	<0.01	100	30	90	0.01		
$ts\,5$	80	125	60	150	30	0.01	
$ts\,6$	50	80	40	70	80	0.3	0.2

TABLE 16-15

	$am\,1$	$ts\,1$	$ts\,2$	$ts\,3$	$ts\,5$
$am\,1$	$<10^{-4}$				
$ts\,1$	2.5	$<10^{-4}$			
$ts\,2$	2.5	0.1	$<10^{-4}$		
$ts\,3$	5	2.5	2.5	$<10^{-4}$	
$ts\,5$	7.5	5	5	2.5	$<10^{-4}$

(d) Describe how you would resolve the ambiguity in the map, using 3-factor crosses.

●16-32. From an imaginary phage M you have isolated three ts mutants, one amber (am) mutant, and one host range (h^-) mutant (one which can grow on E. coli mutants resistant to wild-type phage M). Two-factor crosses were done between all possible pairs of the mutants, and ts^+am^+, ts^+h^- or am^+h^- recombinants were selected in order to measure the frequency of recombination (see Table 16-16). The figures represent percent selected recombinants.

TABLE 16-16

	h^-	$am\,1$	$ts\,1$	$ts\,2$	$ts\,3$
h^-	—				
$am\,1$	2.5	—			
$ts\,1$	9.5	6.9	—		
$ts\,2$	2.6	0.05	7.0	—	
$ts\,3$	4.6	2.1	5.0	2.0	—

(a) What map is generated from this information? Draw h as your left-most marker. (Map units need not be multiplied by 2.)

(b) To eliminate any ambiguity in this map, three-factor crosses are done:

$am^-h^- \times ts2$, selecting for ts^+am^+, gives 10 percent h^-;

$ts3\,am\,1 \times ts2$, selecting for ts^+, gives 0.5 percent am^+ (wild-type).

Complete the phage genetic map and show your reasoning.

16-33. (a) If a point mutant existed in segment $A5\,c2\,a1$ of the rII region of T4 (see Figure 16-8), with which of the 13 listed deletions (heavy lines) would wild-type recombinants be formed in a cross?

(b) If the mutant were in segment $A5d$?

(c) If the mutation is a deletion in $A5\,c2\,a1$?

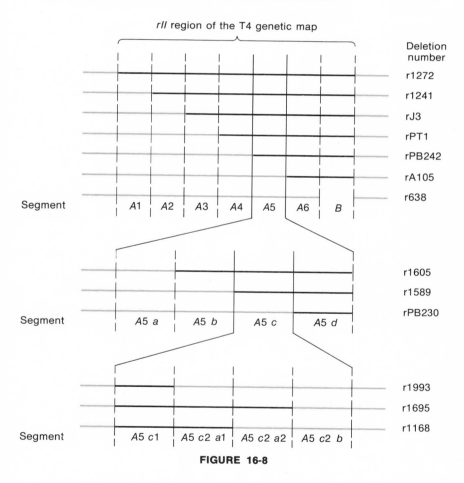

FIGURE 16-8

16-34. A phage T4 mutant in the rII gene is isolated. A revertant of this mutant is isolated that has the wild-type (r^+) phenotype. If this revertant is crossed with authentic r^+ phage and the phages produced in the cross are plated, many plaques are found with the mutant phenotype. Phage stocks are prepared from a large number of these mutant plaques and the stocks are systematically crossed against one another. The plaques are found to fall into two classes A and B defined as follows. If a phage in class A is crossed with a phage in class B, r^+ recombinants result. However in a cross between two phages from the same class, no r^+ recombinants are found. Furthermore, it is found that if the phages in either class A or B are crossed with the original mutant, r^+ recombinants are found only in a cross with a class B phage.

Suppose it were known that the original mutation was a deletion of two bases; what kind of mutation or mutations would be present in the class B phages?

●**16-35.** You have in your strain collection the following bacteria (see Table 16-17).

TABLE 16-17

Strain	Genotype
1	F^- gal^- str-r
2	Hfr gal^+ str-s, transfers gal at 3 min in clockwise direction
3	Same as 2 but also pro^- and deleted for gal operon

(a) Design a protocol to construct a strain in which gal is deleted and which is pro^+str-r.

(b) Suppose you wanted this strain to be str-s instead. Why is this difficult to do with these strains? Design a protocol for accomplishing this. You may alter the three strains by mutation if desired.

●**16-36.** The phage mutants listed down the left side of Table 16-18 have the properties shown in the table when plated on each bacterial strain listed across the top of the table.

Using these phages and bacteria, describe how to construct the following phages. (Consult the phage λ gene map in Appendix E, page 242, if necessary; and see the explanatory notes beneath Table 16-18.)

(a) bio11 cI857.

(b) bio11 cI857 P^-.

(c) c^+ int^- P^-.

(d) cI857 red^- gam^- P^-.

(e) c^+ P^-.

(f) cI857 red^- P^-.

16-37. How would you go about generating a strain carrying the F$'$ gal sex factor if you had a variety of Hfr strains and an F^- recipient?

TABLE 16-18

E. coli strains with relevant genotype

Phage*	C600, sup^+ $recA^+$	594 sup^- $recA^+$	QR48 sup^+ $recA^-$	Q5175 sup^+ $recA^+$ P2 lysogen	152 sup^+ $recA^-$	Q5151 sup^+ $groP^-$
c^+	+	+	+	−	+	−
$cI857$	cl-42 tu-32	cl-42 tu-32	+	−	cl-42 tu-32	−
c^+ $bio11$	+	+	−	+	−	−
c^+ red^-	+	+	+	−	+	−
$cI857$ gam^-	+	+	+	−	+	−
$cI857$ red^- gam^-	+	+	+	+ poorly	−	−
$cI857$ P^-	+	−	+	−	−	+
$cI857$ int^-	+	+	+	−	+	−

Abbreviations: cl, clear plaque; sup^+, carries a suppressor; sup^-, carries no suppressor; tu, turbid plaque; +, forms a plaque; −, does not form a plaque; 32, 42, temperature at which plates are incubated.

Note: A P2 lysogen supports growth only of phages containing the $bio11$ substitution. No other mutant combinations (using mutants in this table) can grow on this strain. The only strain that allows P^- mutants to grow is $groP^-$; other mutations do not interfere with growth. Hence all − in the table are a result of the P^+ genotype.

* The mutations red^-, gam^-, and P^- are suppressor-sensitive.

16-38. Replication of unintegrated F, but not integrated F, is inhibited by exposing *E. coli* to acridine orange. Make use of this finding
(a) to obtain F^- from F^+ cells.
(b) to identify colonies as F^+, Hfr, or F^-.

●16-39. How would you combine the use of integrative suppression and of other techniques to generate Hfr strains with *unique* origins of transfer of chromosomal markers? See *Europ. J. Biochem.* (1974), **43:** 125 for information helpful in solving this problem.

●16-40. Discuss the steps you would take to prepare a P22 specialized transducing phage carrying the *lac* structural gene [see J. Roth, *Genetics* (1974), **76:** 655.]

●16-41. The phage λ mutations given below have the indicated phenotype:
c^+, turbid plaque;
cI 60, clear plaque;
cI 857, turbid at 30° C, clear at 42° C.
bio 11, unable to plate on a *recA⁻* bacterium;
H^-, unable to plate on a bacterium lacking a suppressor, for example, on 594;
H^+, wild-type for *H* gene;
xis 1, unable to excise a prophage.
E. coli C600 plates all of the above and any combination thereof; it contains suppressor II.
Suppose you have in your collection c^+, *cI* 60, *cI* 857, *bio* 11 $H^- c^+$, *cI* 857 *xis* 1, and *bio* 11 c^+.
Write a protocol for making λ*cI* 857 *xis* 1 H^- and testing the phage for each character.

●16-42. Refer to the *E. coli* genetic map in Appendix D, on page 241. You have in your strain collection the following bacteria:
1. F^- *metE⁻ thy⁻ thr⁻ str-r* (resistant to streptomycin).
2. F^-polA^-.
3. Hfr A (transfers clockwise from minute 12) *polA⁻*.
4. Hfr B (transfers clockwise from minute 6) *polA⁻*.
5. Hfr B *leu⁻*.
6. Phage P1.
7. Phage T6.
Write a protocol for making a strain that is F^-*leu⁻polA⁻thy⁻* in which the *str* allele can be + or −. Markers not indicated are to be considered +.

16-43. *E. coli* bacteria move toward a supply of nutrients. This is called **chemotaxis.** It is usually detected by movement of the bacteria through agar containing a concentration gradient of a particular nutrient in the direction of increasing concentration.

A student isolates chemotaxis-defective mutants of *E. coli his⁻*, by selecting the cells that remain at the origin on a galactose-serine-glucose gradient, soft-agar plate as shown in Figure 16-9. The shaded area represents cells remaining at the origin and the

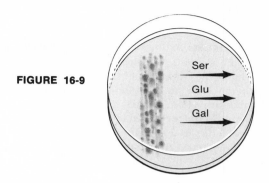

FIGURE 16-9

horizontal arrows point in the direction of increasing concentration of each nutrient in the gradient.

When the student examines cells from 5 mutant clones with a microscope, she finds that cells from clones 1, 2, 3, and 4 cannot swim at all, whereas cells from mutant clone 5 are motile. However, the bacteria of clone 5 do not respond to any known class of repellants or attractants being used in the study and hence are classified as general chemotactic mutants. Of the others, only clone 2 can be suppressed to wild-type chemotactic ability (motility) by amber suppressors. Many other known chemotactic mutants map in the *his uvrC* region of the *E. coli* chromosome, because introduction of an F' carrying *uvrC*+ and *his*+ convert these mutants to wild-type chemotactic ability.

The student proceeds to map these five mutants by P1 transductions, selecting for recombinants with normal chemotaxis. The data are as shown in Table 16-19, when expressed as percent cotransduction with the selective marker *uvr*+. The mapping is done by growing P1 phage on the *uvr*+, nonchemotactic mutant, and using this phage stock to infect a *uvr*− recipient, which is then treated with ultraviolet irradiation to select for *uvr*+ transductants. The transductants are then tested individually for the ability to swim up a nutrient gradient in a capillary tube.

TABLE 16-19

Mutant	Percent cotransduction with *uvr*+
1	53
2	71
3	60
4	63
5	84

In the region of the *E. coli* map that carries genotypes for chemotactic ability, the gene order *his uvrC argS* has been established. All the chemotactic mutants studied here are cotransducible with *his*, but not with *argS*.

The mutants are further mapped by three-factor transductions, where the donor bacterium is *uvrC⁺* and carries one of the chemotactic mutants, and the recipient is *uvrC⁻* and carries a different chemotactic mutant. Transductants with wild-type chemotactic recombinants appear, and their presence is charted (see Table 16-20; + signifies presence and 0 signifies that no wild-type chemotactic recombinants could be detected).

TABLE 16-20

	Clone				
	1	2	3	4	5
1	0	+	+	+	+
2		0	+	0	+
3			0	0	+
4				0	+
5					0

(a) Draw a genetic map of the markers defining clones 1–5.

The student then does a complementation analysis using F' plasmids to make merodiploid strains. Table 16-21 indicates which merodiploid strains have wild-type chemotactic ability (+) and which do not (−).

TABLE 16-21

	Clone				
	1	2	3	4	5
1	−	−	+	−	+
2		−	−	−	+
3			−	−	+
4				−	+
5					−

(b) Draw in gene boundaries on your map; assume there is no intracistronic complementation.

(c) Do the mutants that recombined in the three-factor transductions necessarily complement each other? If not, what is going on?

●16-44 (a) Describe the break–rejoin and copy–choice models of genetic recombination.

(b) In a phage cross between $a^+b^+c^-$ and $a^-b^-c^+$, where the gene order is abc, it was found by Cyrus Levinthal that those progeny phages which were heterozygous for the b locus were always recombinant for the outside markers. Could this be taken to support the copy–choice or break–rejoin models?

(c) It is found that in a phage cross between a^+b^- and a^-b^+, in a mass lysate, there are equal numbers of the reciprocal recombinants a^+b^+ and a^-b^-. However, in a single burst (that is, among the progeny of a single infected bacterium), one recombinant type is frequently absent so that one says that recombination is not reciprocal within a single infected bacterium. The types seen with equal frequency are only seen in counts averaged over a very large number of bacteria. Interpret this finding in terms of break–join versus copy–choice models.

○16-45. Which of the following are examples of reciprocal exchanges in genetic recombination? Prophage integration; prophage excision; formation of recombinants in an Hfr $\times F^-$ bacterial cross; generalized transduction; formation of a homogenote in specialized transduction; formation of a heterogenote in specialized transduction; formation of an F′ factor from an Hfr bacterium; formation of an Hfr bacterium from an F^+ cell; bacterial transformation.

●16-46. Suppose you are transforming a leu^- bacterium with DNA from a leu^+ donor bacterium. Transformants are selected by growing colonies on agar containing leucine (Leu$^+$) and then replica-plating colonies onto agar which lacks leucine (Leu$^-$). The positions of colonies on the Leu$^-$ agar are noted and the corresponding colony is picked from the original plate. These colonies are then dispersed and individual subcolonies from each colony are tested for the leucine requirement. Suppose it is found that, in general, each colony consists of nearly equal numbers of leu^+ and leu^- subcolonies. What mechanisms of recombination would be consistent with this finding? What trivial properties of the recipient bacteria might also explain this result? What would you expect if the original plating was on Leu$^-$ agar?

●16-47. If the break–join model of genetic recombination were correct, how would the frequency of heterozygotes for a particular marker in phage λ depend upon whether or not an inhibitor of DNA synthesis (such as FUDR, fluorouracil deoxyriboside) is added?

●16-48. Read "A general model for genetic recombination," by M. Meselson, and C. Radding, *Proc. Nat. Acad. Sci.*, *U.S.A* (1975), **72:** 358–361. Radding later did an experiment in which both covalent and nicked circles of the replicative form of the phage ϕX174 were separately mixed with single-stranded fragments obtained from the DNA of the phage. What results would Radding expect to find, considering the known properties of supercoiled DNA?

●16-49. The following experiment was done in the laboratory of F. W. Stahl. A λ phage with genotype *red⁻gam⁻A⁻* was density labeled with ¹³C and ¹⁵N and was mixed with λ*red⁻gam⁻R⁻* containing ¹²C and ¹⁴N. The mixture was used to infect *E. coli* under conditions where DNA replication was inhibited and where each bacterium was infected by 10 phages. The phages which were produced were centrifuged to equilibrium in Cs formate and the distribution of genotypes throughout the density gradient was determined. The distribution shown in Figure 16-10 was observed.

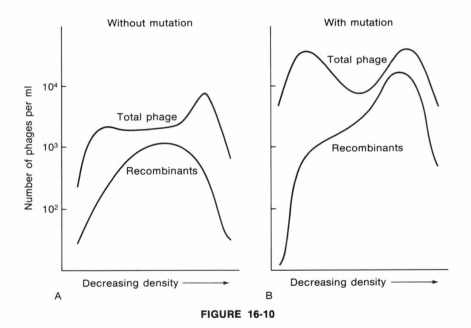

FIGURE 16-10

In a second experiment both phages contained a plaque morphology mutation known to map approximately 93 percent from the left and of the phage and the distribution shown in Figure 16-10 was observed.

Explain both results. (The genetic map of phage λ is on page 242.)

●**16-50.** Two homologous DNA structures are shown diagrammatically. In each molecule a + strand is complementary to its own − strand and to the − strand of the other molecule. The molecules are oriented so that the *b* ends are homologous in base sequence. The upper molecule is shorter than the lower and lacks the *a* end. (See Figure 16-11.) The molecules are allowed to hybridize and to *branch-migrate.* Which of configurations (1) to (6) could result from the operation of these two processes?

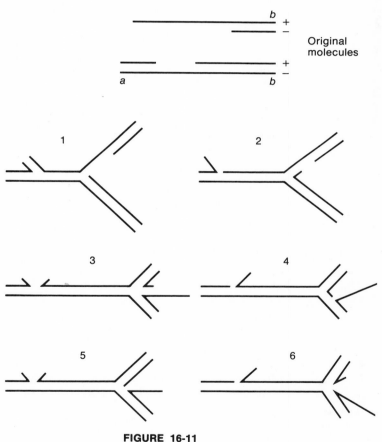

FIGURE 16-11

17

Eukaryotic Cell and Animal Viruses

Introduction

This chapter is a potpourri of problems from cell biology and virology and has been included for use in courses in molecular biology and biochemistry that cover these subjects only briefly.

The fundamental molecular properties of all bacterial species referred to earlier in this book are uniform. That is, the mechanisms of such basic processes as DNA and RNA replication and protein synthesis, and the means of regulating these processes differ only in minor details, when they differ at all. Faith in the principle of evolution leads one to expect only small changes as one proceeds slowly up the evolutionary scale; however it is not surprising that as cells become larger and more complex, the molecular mechanisms present in bacteria do not persist without being somewhat altered. For example, the increased amount of DNA in eukaryotes necessitates having units of replication smaller than the total DNA content of a cell. Thus eukaryotic cells contain numerous chromosomes and the DNA in each chromosome has multiple replication forks. On the other hand, the replication enzyme in eukaryotes has the basic properties of *E. coli* DNA polymerase III; this is also true of other enzymes and protein factors participating in the replication process. Protein synthesis in eukaryotes is quite similar to that in *E. coli* except that eukaryote ribosomes are larger than those in *E. coli*. The separation of the nucleus (where mRNA is made) from the cytoplasm (where mRNA is used in protein synthesis) requires special mechanisms for RNA transport and processing that are not required in bacteria. Furthermore, eukaryotes often have several different RNA polymerases, although we do not yet know why; perhaps so many different types of promoters are needed in a complex cell that a single type of RNA polymerase molecule would not be able to recognize all of them.

Organisms are commonly divided into prokaryotes and eukaryotes, the classification depending upon the absence or presence, respectively, of a nuclear membrane. Interestingly, some of the properties of the molecular processes also are distinguishable in the two classes. We may make another division based instead upon the growth habit of the cells—that is, whether they grow as single isolated cells or as organized units. Single-cell organisms such as bacteria and many algae are essentially immortal; they grow indefinitely, dividing when they reach a critical size. However, yeast cells, which are thought to be more complicated than bacteria, may depart from this pattern, since there is some evidence that the accumulation of bud scars on the surface of a yeast cell that multiplies by budding ultimately alters the cell wall sufficiently that no further growth is possible. On the other hand, mammals, as well as most higher animals, grow to a maximum size at which point cell growth no longer occurs. Related to this may be an observation, well known to those who culture mammalian cells, that a mammalian cell is mortal in that it can divide only a finite number of times. When a change occurs and this mortality is lost, a *transformed* or *cancer cell* often results. The cancer cell has lost, among other things, the ability to be growth-inhibited when in contact with another cell.

Animal and plant viruses are in many ways simpler than bacteriophages. They are usually encased in relatively simple coats (although some are complex and contain membranous material). They lack the injection mechanism used by phages to insert DNA into a host cell; instead, the virus particle is taken up by the cell by an infolding of the host cell membrane, the nucleic acid then being removed from the particle by a dissolution of the particle coat, a process catalyzed by intracellular enzymes. The viruses have some features in common with phages; for example, they possess specific replicating enzymes, their transcription and translation are tightly regulated, and many viruses can insert DNA into a host chromosome in a way that is similar to the process of lysogeny in bacteria. Most species of phages show rather significant uniformity. Although there are some differences from one phage to another, phages infecting *E. coli* have properties quite similar to those that infect totally unrelated bacteria (for example,

Pseudomonas). Viruses, on the other hand, differ widely and fall into at least fourteen classes. The results of virus infection also differ. Some viruses lyse the host cell, whereas others are extruded without damage to the host. The tumor viruses do not kill the host cell but fundamentally change the growth regulation of the host; unlimited growth then begins, and a tumor results. This is usually accompanied by integration of viral DNA into a host chromosome; when the virus contains only RNA, a viral enzyme, *reverse transcriptase,* synthesizes DNA from the RNA template and this DNA is integrated. Another interesting effect of infection by some viruses is to cause fusion of two cells to form a single cell *(heterokaryon)* having an excessive number of chromosomes.

Listed below are several references that provide either a general introduction to cell biology and virology or specific information about topics taken up in the problems; they will be useful while you solve the problems in this chapter.

REFERENCES

Britten, R. J., and D. A. Kohne. 1970. "Repeated sequences of DNA." *Scientific American,* **April.**

Cold Spring Harbor Laboratory. 1974. *Tumor Viruses.* A two-volume, up-to-date, complete review of the field. Advanced.

Davis, B. D., R. Dulbecco, H. N. Eisen, H. S. Ginsberg, and W. B. Wood. 1973. *Microbiology.* New York. Harper & Row. Chapters 44, 47, 48, and 63.

Dupraw, E. J. 1970. *DNA and Chromosomes.* Holt, Rinehart, and Winston. An introductory text containing numerous untested hypotheses about chromosome structure and function.

Ephrussi, B. 1972. *Hybridization of Somatic Cells.* Princeton, N.J. Princeton University Press. Applications of the cell fusion technique.

Fenner, F., B. R. McAuslan, C. D. Mims, J. Sambrook, and D. O. White. 1974. *The Biology of Animal Viruses.* Academic Press. A thorough, advanced text.

Fraenkel-Conrat, H., and R. R. Wagner, 1974. *Comprehensive Virology.* New York. Plenum Publishing Corp. An encyclopedic set of books.

Harris, H. 1970. *Cell Fusion.* Cambridge, Mass. Harvard University Press. An introductory description by the master of cell fusion.

Harris, H. 1972. *Nucleus and Cytoplasm.* Oxford. Clarendon Press. Applications of the cell fusion technique.

Hood, L. E., J. H. Wilson, and W. B. Wood. 1975. *Molecular Biology of Eukaryotic Cells.* Menlo Park, Calif. W. A. Benjamin. A problems book with informative introductions.

Littlefield, J. W. 1976. *Variation, Senescence, and Neoplasm in Cultured Somatic Cells.* Cambridge, Mass. Harvard University Press. A good discussion of cells in culture.

Luria, S. E., and J. E. Darnell. 1967. *General Virology*. New York. John Wiley & Sons.

Mazia, D. 1974. "The cell cycle." *Scientific American,* **January.**

Mitchison, J. M. 1971. *The Biology of the Cell*. Includes an excellent comparison of eukaryotes and prokaryotes.

Novikoff, A. and E. Holtzmann. 1970. *Cells and Organelles*. New York. Holt, Rinehart, and Winston. An introductory text.

Paul, J. 1970. *Cell and Tissue Culture*. Edinburgh. Churchill Livingstone. A book of methods.

Puck, T. T. 1972. *The Mammalian Cell as a Microorganism*. San Francisco. Holden-Day. A fine introduction to cell culture techniques, the cell cycle, and the thymidine double-block synchronization procedure.

Ruddle, F. H., and R. S. Kucherlapatai. 1974. "Hybrid cells and human genes." *Scientific American,* **July.** The application of cell fusion to the determination of gene location.

Sager, R. 1972. *Cytoplasmic Genes and Organelles*. New York. Academic Press. Details about current thought on chloroplasts and mitochondria.

Taylor, J. H. 1960. "Asynchronous duplication of chromosomes in cultured cells of Chinese hamster." *J. Biophys. Biol. Cytol.,* **7:** 455–463. Describes the techniques for determining the sequence of replication of chromosomes.

Tooze, J. 1973. *The Molecular Biology of the Tumor Viruses*. Cold Spring Harbor, N.Y. Cold Spring Harbor Laboratory. An excellent review of cell culture, virology, and the problem of transformation.

Watson, J. D. 1976. *Molecular Biology of the Gene*. Third Edition. Menlo Park, Calif. W. A. Benjamin. Chapters 16, 17, 18, and 20. An excellent exposition of current ideas in cell biology and the study of tumor viruses.

Wessels, N. K. 1971. "How living cells change shape." *Scientific American,* **October.** A simple article explaining cell structure and movement.

White, M. J. D. 1973. *The Chromosomes*. Sixth Edition. New York. Halsted Press. An introductory book.

Wolfe, S. L. 1972. *Biology of the Cell*. Belmont, Calif. Wadsworth Publishing Co. An introductory text emphasizing chromosome structure and mechanics.

Problems

○**17-1.** Explain the following terms used to describe the cell cycle: G1, G2, M, S.

○**17-2.** What is the chemical composition and the function of: the nucleolus; lysosomes; mitochondria; endoplasmic reticulum; chromatin?

17-3. What is the evidence for iterated DNA sequences?

17-4. Describe some characteristics of eukaryotic mRNA which distinguish it from bacterial mRNA.

17-5. What is meant by heterogeneous nuclear RNA?

17-6. Explain the term contact inhibition or, as it is now more frequently called, density-dependent growth.

17-7. Define the following terms: chromosome; chromatid; centromere; chromatin; nucleosome.

17-8. What is a transformed cell? Distinguish a transformed cell from a normal cell with respect to density-dependent growth, anchorage dependence, serum requirement, lectin agglutination, and plasminogen activator.

○17-9. Explain the Lyon-Russell hypothesis and describe an experiment which proves it. What is a Barr body?

○17-10. What is the quinacrine method for chromosomal banding and what is it good for?

17-11. Describe *in situ* hybridization and some questions which have been answered by this procedure.

17-12. Why does the addition to growing cells of the drugs colchicine, colcemide, vinblastine, or vincristine lead to an increase in mitotic figures in cell culture?

●17-13. The addition of a high concentration of thymidine to a growing culture causes an immediate inhibition of DNA synthesis. Cells in other stages of the cell cycle are uninhibited and proceed through the cycle. There is a technique called the **double thymidine-phasing method,** in which thymidine is added, removed, and readded to a growing culture, with the result that all cells are synchronized with respect to DNA synthesis—that they are all synthesizing DNA simultaneously. When should thymidine be added and for how long? You may consider G1 and S to be of 8 hours duration, and G2 and M to be of 6 and 2 hours duration, respectively.

17-14. How is cell fusion carried out? If a cell with 22 chromosomes is fused to another species of cells with 30 chromosomes, what is the maximum and minimum number of chromosomes found in cells which are hybrid for some particular gene? Roughly what is the mean chromosome number of genetic hybrids? How is this method used in cellular genetics?

17-15. In *E. coli* the basic ribosomal subunits have sedimentation coefficients of 50s and 30s, and these subunits contain 23s and 16s RNA, respectively. What are the corresponding values for mammalian ribosomes?

•17-16. Penicillin, tetracycline, erythromycin, chloromycetin, puromycin, and actinomycin are powerful inhibitors of bacterial growth. However, of these only the first three are used in the treatment of human diseases. Chloromycetin is used only in cases where all else fails. Explain these facts in terms of microbial and mammalian cell biology.

•17-17. A particular cell line has values for M, G1, S, and G2 of 2, 8, 6, and 1 hour, respectively. Each cell has 6 easily distinguishable chromosome pairs which are numbered 1 through 6, in order of increasing size. An amount of colchicine sufficient to cause mitotic arrest, and ^{3}H-thymidine of very high specific activity, are added to a logarithmically growing culture. At one-hour intervals after addition of these components, cells are taken and prepared for autoradiography. After several days' exposure the autoradiograms are developed and for 100 metaphase cells, the particular chromosomes over which there are grains are noted. The data in Table 17-1 are recorded. What is the sequence in time of the synthesis of the DNA for each chromosome?

TABLE 17-1

Time after addition of ^{3}H-dT (in hours)	Of 100 cells, the number in which chromosome pairs 1, 2, . . . , 6 are labeled					
	#1	#2	#3	#4	#5	#6
0	0	0	0	0	0	0
1/2	0	0	0	0	0	0
1	0	0	14	0	0	0
2	25	1	100	0	0	0
3	100	80	100	0	3	0
4	100	100	100	0	100	0
5	100	100	100	100	100	15
6	100	100	100	100	100	100
7	100	100	100	100	100	100

17-18. Define the following terms used in virology: virion; capsid; envelope; core; capsomers; peplomers ("spikes"); cytopathic effect; inclusion bodies.

•17-19. Some viruses contain substantial quantities of phospholipids, cholesterol, and carbohydrates. What is the source of these substances?

17-20. One way in which a virus differs from a bacteriophage is that it does not have an injection apparatus for getting its nucleic acid into the host cell. How does penetration occur?

17-21. Efficiency of plating (EOP) is defined as the fraction of particles that can form a visible plaque. With phages, the EOP is generally nearly one, whereas with viruses, values of from 10^{-6} to 10^{-1} are more common. What factors contribute to the very low values of EOP?

17-22. Describe and explain the cause of hemagglutination.

17-23. What forms of nucleic acid are isolated from polyoma virus and from SV40 virus?

17-24. Describe some evidence which suggests that some viruses can exist intracellularly in a state analogous to the lysogenic state of bacteriophages.

17-25. Some RNA tumor viruses are known to be present in a provirus state within cells, in a manner analogous to the prophage of lysogenic bacteriophages. However, RNA is not found inserted into host DNA. Explain.

17-26. Classify the following viral species as DNA or RNA viruses: enterovirus; parvovirus; leukovirus; reovirus; polyomavirus; rhinovirus; adenovirus; poxvirus; herpes virus.

●17-27. Describe ways in which cell fusion is used to study virus-cell interactions.

17-28. What is meant by the capping of viral mRNA?

Appendix A

The Genetic Code

First position (toward 5' end of strand)	Second position				Third position (toward 3' end of strand)
	U	C	A	G	
	Phe	Ser	Tyr	Cys	U
	Phe	Ser	Tyr	Cys	C
U	Leu	Ser	Stop	Stop	A
	Leu	Ser	Stop	Trp	G
	Leu	Pro	His	Arg	U
	Leu	Pro	His	Arg	C
C	Leu	Pro	Gln	Arg	A
	Leu	Pro	Gln	Arg	G
	Ile	Thr	Asn	Ser	U
	Ile	Thr	Asn	Ser	C
A	Ile	Thr	Lys	Arg	A
	Met*	Thr	Lys	Arg	G
	Val	Ala	Asp	Gly	U
	Val	Ala	Asp	Gly	C
G	Val	Ala	Glu	Gly	A
	Val	Ala	Glu	Gly	G

* AUG is a codon that also signals "start," though not in all cases.

Appendix B

Properties of Selected Nucleases

Enzyme (nuclease)	Substrate	Site of cleavage	Product
Pancreatic ribonuclease	RNA	Endonuclease; adjacent to pyrimidines	Mono- or oligo- nucleoside terminating with pyrimidine nucleoside-3'-P
T1 RNase	RNA	Endonuclease; adjacent to guanosine	Mono- or oligo- nucleoside terminating with guanosine-3'P
Venom phosphodiesterase	RNA or DNA	Exonuclease at 3'-OH end	5'-P mononucleotides
Spleen phosphodiesterase	RNA or DNA	Exonuclease at 5'-OH end	3'-P mononucleotides
E. coli exonuclease I	Single-stranded DNA	Exonuclease at 5'-OH end	5'-P mononucleotides plus a terminal dinucleotide
E. coli exonuclease II	Double-stranded DNA	Exonuclease at 3'-OH end	5'-P mononucleotides
E. coli exonuclease III	Double-stranded DNA	Exonuclease at 3'-OH end	5'-P mononucleotides
E. coli endonuclease I	Double-stranded DNA	Many $3' \rightarrow 5'$-phosphoester sites	Oligonucleotides with 5'-P end

Appendix C

Amino Acid Sequence of Tobacco Mosaic Virus (TMV) Coat Protein

Tobacco mosaic virus infects the leaves of living tobacco plants. It contains a single RNA molecule that is enclosed in a protein coat consisting of hundreds of identical protein molecules arranged in a helix. The amino acid sequence of the coat protein, shown here, was elucidated in 1962 independently by H. Fraenkel-Conrat and H. Wittmann. The sequence was important in understanding the properties of the genetic code, because it showed (1) that a mutant differed from wild-type by only a single amino acid exchange and (2) which amino acid exchanges are possible as a result of a single base change. From knowledge of the chemical changes produced by several mutagens and observation of the amino acid exchanges that resulted from application of particular mutagens, good guesses were made about codon assignments.

Note that the serine in position 1 is acetylated (AcSer). Note also that there are no extended repetitions of an amino acid sequence, that there are no clusters of either polar or nonpolar amino acids, and that the frequencies of occurrence of different amino acids vary widely.

AcSer-Tyr-Ser-Ile-Thr-Thr-Pro-Ser-Gln-Phe-
1   5   10

Val-Phe-Leu-Ser-Ser-Ala-Trp-Ala-Asp-Pro-
15   20

Ile-Glu-Leu-Ile-Asn-Leu-Cys-Thr-Asn-Ala-
25   30

Leu-Gly-Asn-Gln-Phe-Gln-Thr-Gln-Gln-Ala-
35   40

Arg-Thr-Val-Val-Gln-Arg-Gln-Phe-Ser-Glu-
45   50

Val-Trp-Lys-Pro-Ser-Pro-Gln-Val-Thr-Val-
55   60

Arg-Phe-Pro-Asp-Ser-Asp-Phe-Lys-Val-Tyr-
65   70

Arg-Tyr-Asn-Ala-Val-Leu-Asp-Pro-Leu-Val-
75   80

Thr-Ala-Leu-Leu-Gly-Ala-Phe-Asp-Thr-Arg-   *(continued)*
85   90

Asn-Arg-Ile-Ile-Glu-Val-Glu-Asn-Gln-Ala-
$\qquad$95$\qquad\qquad$100

Asn-Pro-Thr-Thr-Ala-Glu-Thr-Leu-Asp-Ala-
$\qquad$105$\qquad\qquad$110

Thr-Arg-Arg-Val-Asp-Asp-Ala-Thr-Val-Ala-
$\qquad$115$\qquad\qquad$120

Ile-Arg-Ser-Ala-Ile-Asn-Asn-Leu-Ile-Val-
$\qquad$125$\qquad\qquad$130

Glu-Leu-Ile-Arg-Gly-Thr-Gly-Ser-Tyr-Asn-
$\qquad$135$\qquad\qquad$140

Arg-Ser-Ser-Phe-Glu-Ser-Ser-Ser-Gly-Leu-
$\qquad$145$\qquad\qquad$150

Val-Trp-Thr-Ser-Gly-Pro-Ala-Thr
$\qquad$155$\qquad$158

The following substitutions have been observed in mutants. (numbers refer to positions of the amino acids): 5, Ile; 8, His; 9, His; 10, Leu; 11, Met; 15, Leu; 19, Val, Ala; 20, Thr, Leu; 21, Thr, Met; 24, Val; 25, Ser; 28, Ile, Ala; 33, Ser, Ala, Lys; 45, Lys, Gly; 55, Leu; 58, Ala; 59, Ile; 61, Gly; 63, Ser; 65, Gly; 73, Ser; 81, Ala; 95, Asp; 97, Gly; 99, Arg; 107, Met; 122, Gly; 125, Val; 126, Ser; 129, Thr, Val; 134, Gly; 136, Ile; 138, Phe; 139, Cys; 140, Lys; 148, Phe; 149, Ser; 153, Ile; 156, Leu.

Appendix D

Genetic Map of *E. coli,* Showing Relative Positions of Selected Markers

Abbreviations (clockwise from top): *thr,* threonine; *leu,* leucine; *lac,* lactose; *lon,* a repair gene; *gal,* galactose; *att*λ, prophage attachment site for phage λ; *bio,* biotin; *trp,* tryptophan; *his,* histidine; *pur,* purine; *recA,* a recombination gene; *ser,* serine; *thy,* thymine; *fda,* aldolase; *str,* streptomycin; *metE,* methionine E gene; *argC,* arginine C gene; *malB,* maltose B gene; *uvrA,* gene for excision repair. Locations of other genes can be estimated by using the reference points indicated for key genes (map is divided into 90 minutes).

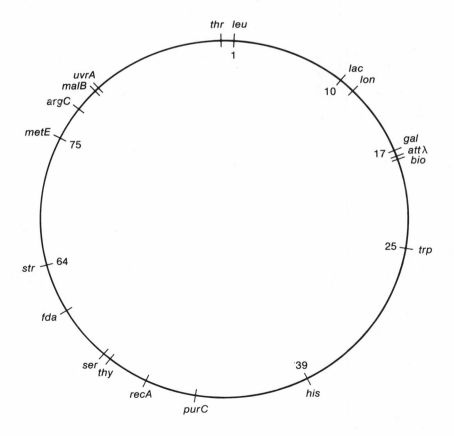

Appendix E

Genetic and Physical Map of *E. coli* Phage λ

The heavy line represents the DNA molecule present in the phage head. Genes are indicated in the correct position on the DNA. An expanded view of the *att–cI* region is shown. Abbreviations: *m, m'*, the single-stranded cohesive ends which, when joined, form the *cos* site; *A*, the gene for producing *m* and *m'* from a *cos* site; *E*, gene for the major head protein; *K* and *J*, genes for tail proteins; *b*2, a nonessential region which is frequently deleted; *att*, the prophage attachment site, also referred to as *PP'*; *int*, integrase; *xis*, excisionase; *red*, a locus consisting of two genes for genetic recombination; *gam*, a gene responsible for inhibiting the *E. coli* RecBC protein; *cIII*, a positive regulator for synthesis of both integrase and repressor; t_{L1}, the first leftward terminator for transcription from p_L; *N*, a gene whose product allows t_{L1} to be ignored; p_L, the leftward promoter; o_L, the leftward operator; *cI*, repressor; *O* and *P*, two genes needed for initiation and propagation of DNA synthesis; *Q*, a gene required to initiate transcription of the *SRAEKJ* region; *S* and *R*, genes involved in lysis.

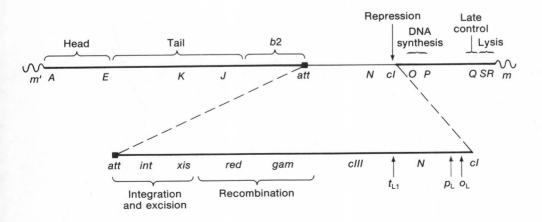

Appendix F

Abbreviations

A	adenine
ADP	adenosine diphosphate
Ala	alanine
am	amber
AMP	adenosine monophosphate
Arg	arginine
Asp	aspartic acid
Asn	asparagine
ATP	adenosine triphosphate
B	5-bromouracil
Bio	biotin
C	cytosine, carbon
^{14}C	carbon-14 isotope
cAMP	cyclic AMP
CAP	cyclic AMP receptor protein
CDP	cytidine diphosphate
Cys	cysteine
dAMP	deoxyadenylic acid
dCTP	deoxycytosine triphosphate
dCTPase	enzyme converting dCTP to deoxycytidylic acid
DFP	diisopropylfluorophosphate
dHTP	deoxyhydroxymethylcytosine triphosphate
dUTP	deoxyuridine triphosphate
dX	deoxy form of base X
F^-	female *E. coli*
F^+	male *E. coli*
G	guanine
Gal	galactose
Gln	glutamine

Glu	glutamic acid
GTP	guanosine triphosphate
H	heavy strand of single-stranded DNA
HH	double-stranded DNA molecule in which both strands have a high density caused by a density label
^{3}H	tritium
^{3}H-dT	tritiated thymidine
Hfr	male *E. coli* with an integrated F factor
His	histidine
HMC	5-hydroxymethylcytosine
i	*lac* operon repressor
Ile	isoleucine
Int	λ integrase
IS2	insertion sequence #2
L	Light strand of DNA
LL	double-stranded DNA in which both strands have low density
Lac	lactose
λ*dg*	gal-transducing λ phage
Leu	leucine
Lys	lysine
Mal	maltose
Met	methionine
moi	multiplicity of infection
o	operator
p	promoter
P	organic phosphate
^{32}P	phosphorous-32 isotope
Phe	phenylalanine
pho	phosphatase gene, consisting of cistrons *pho*A, *pho*B, . . .
pol	polymerase
pol I	DNA polymerase I
polyX	polynucleotide containing base X only
polyXY	polynucleotide containing bases X and Y
poly(X:Y)	double-stranded polynucleotide consisting of one strand of polyX and one of polyY
Pro	proline
proA, proB	particular genes in the *pro* operon
Pur	purine
Pyr	pyrimidine
*p*X	5'-P of nucleotide X
Rec	bacterial recombination gene
Red	λ recombination gene
RF	replicative form

ρ	density
rho factor	termination factor for RNA synthesis
RNase	ribonuclease
RTF	resistance transfer factor
rRNA	ribosomal RNA
rX	ribose form of base X
s	sedimentation coefficient
$s_{20,w}$	sedimentation coefficient corrected to 20° C and to a solvent of pure water
SAMase	S-adenosylmethioninase
Ser	serine
Str	streptomycin
str-r, str-s	resistance and sensitivity to streptomycin
sup	suppressor gene
supII	a particular suppressor gene
T	thymine
Tet	tetracycline
tet-r, tet-s	resistance and sensitivity to tetracycline
Thi	thiamine, vitamin B1
ts	temperature-sensitive
tsx	gene for the receptor site for *E. coli* phage T6
U	uracil

SPECIAL NOTATION

Names of genes or abbreviations referring to genotypic components are written in lower-case italic letters; names referring to substances added to growth media and components of polymers, and names describing phenotypic qualities, are not italicized and the first letter of each is usually capitalized.

Superscripts $^+$ and $^-$ refer to active and inactive forms (alleles) of genes, respectively.

Numbers following genes indicate mutation number and are not italicized.

Glossary

absorbance A measure of the decrease in light transmitted by a sample. Defined as $-\log_e(I/I_0)$ in which I and I_0 are the transmitted intensity and the incident intensity, respectively.

active site The region of a protein in which interaction with another molecule takes place.

allele One or two different forms of a gene; for example, lac^+ and lac^- are wild-type and mutant alleles, respectively, of the lac gene.

amber codon The chain-termination codon UAG.

amber mutation A mutation, the result of which has been generation of a UAG triplet.

amber suppressor A tRNA having the anticodon AUC and hence capable of inserting an amino acid at a UAG site.

amino acids The monomers of which proteins or polypeptides are constructed. All biological amino acids except proline have the fundamental structure

$$^+H_3N-\overset{\overset{\displaystyle H}{|}}{\underset{\underset{\displaystyle R}{|}}{C}}-C\overset{\displaystyle O}{\underset{\displaystyle O^-}{\diagup\diagdown}}$$

but differ in the side group R.

amino terminus The amino acid in a protein or polypeptide in which the carboxyl group but not the amino acid is in a peptide bond.

anticodon The group of three adjacent bases in a tRNA molecule that recognizes and pairs with a messenger RNA codon.

antiserum Animal blood freed of all cells and containing antibody to at least one substance X. Such an antiserum is frequently denoted X antiserum or anti-X serum and is said to contain anti-X or anti-X antibody.

auxotroph An organism which cannot grow in a growth medium consisting of inorganic salts and an organic carbon source but which requires at least one organic compound such as an amino acid, vitamin, fatty acid, and so on.

base analogs Purines or pyrimidines which differ slightly in structure from the usual DNA bases and which can be incorporated into nucleic acid in place of a normal base.

β-galactosidase The product of gene z of the lac operon catalyzing the cleavage of lactose to glucose and galactose.

bidirectional replication DNA replication in which a single molecule has two replication forks moving in opposite directions.

broth A complex, organic, growth medium for bacteria or other microorganisms; it is commonly derived from meat, blood, milk, or, less commonly, certain plants.

burst size The number or average number of phages or viruses produced by an infected cell.

carboxyl terminus The amino acid in a protein or polypeptide in which the amino group but not the carboxyl group is in a peptide bond.

catenane A structure consisting of two circles, usually DNA, linked as in a chain.

cis-dominant A mutant which only affects expression of genes in the DNA or chromosome containing the mutation. Mutations in promoters and operators are examples of *cis*-dominant mutants.

cistron A sequence of bases in DNA coding for a single polypeptide chain.

clone A group of cells each descended from a single common cell.

closed circle A double-stranded, circular DNA molecule containing no interruptions in either strand.

codon Three adjacent bases in mRNA that can be translated into an amino acid or serve as a chain termination signal.

colony A group of cells, in contact, derived from a single cell and grown on a solid surface.

complementation The production of a wild-type phenotype when two mutants are combined as in a diploid or by infection of a host cell with two mutant phages.

concatemer A linear polymer consisting of a repeating unit of a polymer; usually refers to DNA molecules of greater than unit length.

covalent circle *See* closed circle: they are the same.

cross A genetic cross; the mating of two organisms whose genotypes are such that genetic recombinants can be detected.

cyclic permutation Cyclically permuted DNA refers to a population of DNA molecules, all of which have the same number of nucleotides but whose nucleotide sequences are permutations of one another.

dalton A unit of mass equal to that of a hydrogen atom.

deletion Loss of a section of a chromosome or a DNA molecule. The size of the deleted material may be as small as a single nucleotide.

denaturation Loss of the natural configuration of a polymer. With respect to DNA it means separation of base pairs and ultimately of polynucleotide strands. With respect to proteins it means the loss of all interactions between amino acids.

denaturation map DNA can be partially denatured under certain conditions, to produce a molecule containing single-stranded loops. For a particular set of conditions, the positions of the loops with respect to physical density along the molecule constitute a denaturation map.

diploid Having two copies of each chromosome; the copies may differ by the presence of mutations.

DNA ligase *See* polynucleotide ligase.

DNA-RNA hybrid A double-stranded molecule consisting of one strand of DNA hydrogen bonded by complementary base pairing to an RNA strand.

dominant An allele whose phenotype is expressed, whether in homozygous or heterozygous form.

early gene In a phage infection, a gene which is transcribed very soon after injection, the transcription of which is usually shut off before or shortly after the late genes are transcribed.

endonuclease An enzyme capable of breaking phosphodiester bonds in nucleic acid but not having any preference for terminal bonds (*see* exonuclease). A particular endonuclease usually acts on either DNA or RNA but not both, may have preference for bonds in single-stranded DNA, double-stranded DNA, or between certain bases, and acts on either the 3' or 5' side of the phosphate, but not both.

ethidium bromide A fluorescent dye which intercalates between the base pairs of double-stranded DNA.

eukaryote A cell whose nucleus is surrounded by a membrane.

exonuclease *See* endonuclease. An exonuclease can only cleave a phosphodiester bond connecting a terminal nucleotide to the adjacent nucleotide.

F factor A circular DNA molecule which is present in male strains of *E. coli* and contains the genes for pairing male and female cells and DNA transfer; other genes are also present. The F factor is inserted into the chromosome in an Hfr male.

fingerprint A pattern of peptides or oligonucleotides obtained by degradation (usually enzymatic) of a protein or nucleic acid, respectively, followed by two-dimensional chromatography or chromatography-electrophoresis.

frameshift mutation The deletion or addition of a small number of bases (usually 1, 2, 4, or 5) situated that the reading frame of the codon triplets is altered.

generalized transducing phage A phage capable of carrying probably all regions of the genome of the host bacterium.

genetic map The arrangement of sites on a chromosome deduced from genetic recombination data.

genome The haploid set of chromosomes with their genes; the term is often incorrectly used to refer to the DNA content of a cell or to a DNA molecule.

genotype The genetic constitution of an organism.

haploid Having one set of chromosomes.

heavy isotope An isotope whose atomic weight is greater than the isotope having greatest natural abundance. In biochemical terminology, heavy isotope never refers to a radioactive isotope. Examples of heavy isotopes used in biology are ^{2}H, ^{13}C, and ^{15}N.

Hershey circle A circular DNA formed by base pairing between complementary, terminal, single-stranded regions of a double-stranded DNA.

heteroduplex A DNA molecule formed by base pairing between single strands of two DNA molecules whose base sequences are not totally complementary.

heterogenote A transductant containing two copies (usually the + and − alleles) of the gene being examined.

heterozygote (a) An organism containing either two different alleles of a single gene on homologous chromosomes of a diploid set. (b) A DNA molecule containing two alleles of a single gene in the molecule; this can involve gene duplication, or the two strands of a single gene can differ at one (usually) or more loci.

Hfr A male *E. coli* bacterium containing an integrated F factor and able to transfer its chromosome to a female (the abbreviation stands for "high frequency of recombination").

homogeneous (DNA) A population of DNA molecules, each of which has the same molecular weight.

homogenote A transductant containing a single copy of the gene being examined.

homozygote A diploid organism containing two copies of the same allele of a particular gene.

host cell A cell infected with a phage or virus.

H strand The single strand of DNA having higher density when the strands are separated under particular conditions.

hybrid DNA Usually, a DNA molecule (a) whose strands carry different density labels or (b) whose strands are derived from two different DNA molecules by successive denaturation and renaturation.

hybridization The process of forming hybrid DNA of the type defined at (b) under hybrid DNA.

in vitro Using cell extracts or components.

in vivo Using intact cells only.

infection The process of adsorbing a phage or virus to a host cell, usually followed by replication of the particles or lysogenization.

infectious center A cell infected with a phage or virus.

infectious DNA DNA isolated from a phage or virus which can successfully enter a host cell and initiate development of the particle.

insertion Integration of phage or viral DNA into the host DNA to make a continuous piece.

Kleinschmidt method A method of sample preparation for electron microscopy in which (1) nucleic acid molecules are mixed with cytochrome *c*, (2) a film of cytochrome *c* containing nucleic acids is formed on a water surface, and (3) the film is picked up onto a microscope grid.

late genes In a phage or viral infection, genes which are not transcribed immediately after injection but usually after other sets of genes (*see* early genes) are transcribed. Usually, late genes code mainly for structural proteins and lysis enzymes.

leaky mutant A mutant which partially expresses its phenotype. Usually these are not of value for genetic or biochemical analysis.

locus Any site on a DNA molecule or chromosome.

ligase *See* polynucleotide ligase.

l-strand A DNA strand which is transcribed from right to left when the genetic map is drawn in a standard way.

L-strand The single strand of DNA having lower density when the strands are separated under particular conditions.

lysis Rupture of a cell and the release of its contents.

lysogen A bacterium containing an integrated phage DNA molecule.

melting Denaturation by heat.

melting curve A graph relating some physical or biological property of a macromolecule as a function of temperature.

melting temperature The temperature at which the transition of a melting curve is half complete; usually written T_m.

minimal medium A growth medium for microorganisms in which all components other than the carbon source are inorganic compounds; for the growth of auxotrophs, other organic substances are added.

missense mutation A mutation resulting in the replacement of one amino acid by a second amino acid.

missense suppressor An altered tRNA or amino acyl synthetase which results in occasional substitution of one amino acid by a second amino acid.

mutagen A substance which causes mutations to occur.

mutation A change in a nucleotide base such that one codon is converted to another codon. If the codon change does not cause a detectable phenotypic change, the mutation is said to be *silent*.

mutator An altered polymerase or repair system which results in frequent base substitutions in DNA.

multiplicity of infection The average number of phages or viruses adsorbed per host cell.

nearest neighbors Adjacent nucleotides in a nucleic acid chain.

negative control A means of regulation of gene activity in which a substance is present when transcription is turned off and removed or inactivated when transcription is to occur.

nick A single-strand break in a double-stranded polynucleotide.

nonpermissive A host cell which does not allow successful development of an infecting phage or virus.

nonsense codon A codon for which there does not normally exist a tRNA molecule; involved in chain termination in protein synthesis. The codons are UAG, UAA, and UGA.

nonsense mutant A mutant in which a normal codon has been altered to form a nonsense codon.

nonsense suppressor A tRNA which has been altered so that it has an anticodon corresponding to one of the chain termination codons UAG, UAA, or UGA.

nuclease An enzyme capable of hydrolyzing phosphodiester bonds in nucleic acid. *See* endonuclease; exonuclease.

OD Optical density. *See* absorbance.

oligonucleotide A relatively short polymer consisting of nucleotides joined by standard phosphodiester bonds.

open circle A circular DNA molecule containing at least one single-strand break.

operator A site on a DNA molecule at which a repressor binds.

operon A coordinately regulated set of cistrons and binding sites.

optical density *See* absorbance.

partial denaturation Usually refers to a process by which regions of a DNA molecule are converted to stable single-stranded loops.

partial diploid A bacterium which contains a sex factor carrying copies of genes also present in the bacterial chromosome.

permissive A host cell enabling certain phages or viruses or mutants thereof to grow which cannot grow on all host cells; *see* nonpermissive.

phage cross *See* cross.

phenotype The observable properties of an organism.

plaque A clear or partially clear area in a region of confluent growth of a bacterium produced by a single phage particle.

plasmid A circular DNA molecule contained in a bacterium. A plasmid may or may not carry genes present in the bacterial chromosome. Plasmids usually carry at least one gene necessary for its own replication and not involved in replication of the bacterial chromosome. Plasmids are almost always dispensable and are sometimes transferable.

point mutation A mutation in which a single base is altered but not deleted.

polar mutation A mutation that affects not only the cistron in which it is contained but also decreases the activity of cistrons which are further from the promoter in the same operon.

polycistronic mRNA A messenger RNA containing the information for synthesis of at least two proteins.

polynucleotide A polymer consisting of nucleotides joined by phosphodiester bonds.

polynucleotide kinase An enzyme that transfers the γ-phosphoryl of adenosine triphosphate to a 5'-OH group of a polynucleotide.

polynucleotide ligase An enzyme which can join two adjacent nucleotides bearing a 5'-P and a 3'-OH group respectively in a single strand of a double-stranded polynucleotide.

polyribosome, polysome Several ribosomes bound to a single mRNA molecule.

positive control Regulation of an operon by a protein the expression of which is necessary for transcription of the DNA of the operon.

prokaryote A cell whose nucleus is not bounded by a membrane (*see* eukaryote).

promoter A region of a DNA molecule at which RNA polymerase binds and initiates transcription.

prophage A piece of phage DNA contained in a bacterial chromosome.

prototroph A microorganism having no nutritional requirements (*see* auxotroph).

provirus A piece of viral DNA or a DNA copy of viral RNA contained in a eukaryote chromosome.

^{32}P-suicide Killing of a microorganism by decay of ^{32}P in the DNA.

recessive An allele whose phenotype is expressed only when in homozygous form.

recognition site Usually, a site on a nucleic acid to which a specific protein or nucleic acid can be bound.

recombination The assortment of genetic characters to produce DNA molecules having a new genotype.

regulatory gene A gene whose product affects the expression of another gene, either by means of transcription or translation of the gene or by interaction with the gene product of the second gene.

renaturation The restoration of the natural structure of a macromolecule; in some cases the original activity but not the original structure is all that is observed. With DNA it refers to the reformation of double-stranded DNA from single strands.

repair The process of correcting errors in a DNA molecule; the errors may be unpairable bases, altered bases, missing bases, or broken phosphodiester bonds.

replication fork The region in which polymerization of nucleotides occurs in a double-stranded polynucleotide.

replicative form (RF) A DNA structure which is different from that found in a phage or virus particle and which is necessary for replication to occur; e.g., the double-stranded circle found in the replication of single-stranded DNA phages.

repressor A protein which binds to an operator and thereby prevents transcription.

restriction enzyme An endonuclease

which cleaves double-stranded DNA at a unique base sequence.

reversion The process of correcting a mutation by restoring the original base sequence or by introducing a second mutation.

revertant A microorganism in which reversion has occurred.

rho **factor** A protein involved in termination of transcription at particular sites.

ribosome A particle consisting of RNA and proteins on which protein synthesis occurs.

rolling circle A DNA molecule replicating by the rolling circle mode in which a double-stranded branch is attached to a circle.

rRNA Ribosomal RNA, the RNA contained in a ribosome.

r-strand A DNA strand from which transcription occurs in the rightward direction when the genetic map is drawn in a standard way.

s, sedimentation coefficient In a centrifugal field, the ratio of the velocity of a particle to the centrifugal force; expressed in svedbergs.

sex-linked character A gene carried on a sex chromosome of a eukaryote.

σ **(sigma) factor** One of the protein subunits of RNA polymerase; responsible for recognition of start signals.

specialized transducing phage A phage carrying a unique piece of bacterial DNA—*see* generalized transducing phage.

spontaneous mutant A mutant which arises without exposure of the organism to a known mutagen.

sticky ends Complementary single-stranded termini of double-stranded DNA molecules—for example, the DNA of *E. coli* phage λ.

supercoil A twisted covalent circle; also called a superhelix.

suppressor See missense and nonsense suppressor.

suppressor-sensitive mutant A mutant which has the wild-type phenotype when a suppressor is present; often designated *sus*.

sus *See* suppressor-sensitive mutant.

temperature-sensitive mutant A mutant which has a wild-type phenotype only above (cold-sensitive) or below (heat-sensitive) a particular temperature.

terminally redundant DNA A DNA molecule which has a common base sequence at both ends of the molecule, for example, ABCD . . . XYZABC.

thymine dimer A cyclobutane-type dimer formed by ultraviolet light between adjacent thymines on a single-strand of DNA.

time-of-entry curve In a mating between Hfr and female bacteria in which the bacterial pairs are separated at various times after mixing, a curve in which the number of recombinants is plotted as a function of time; the time of entry is the point at which back-extrapolation of the curve intersects the time axis.

T_m *See* melting temperature.

wild-type Presumably, the genotype found in nature; however, very commonly the term refers to the + form of an allele even when the − form is that most commonly found in nature.

Answers

Chapter 1

1-1. Prokaryotes and eukaryotes: cells that, respectively, lack and possess nuclear membranes. Nucleolus: round structure in eukaryotic nuclei that is involved in ribosomal RNA synthesis and ribosome formation. Prophase, metaphase, anaphase, telophase: four stages of mitosis—condensation, alignment, separation, and disorganization of chromosomes. Karyotype: the chromosome complement, i.e., the number and shapes of the chromosomes. Haploid and diploid: having one and two sets of distinct chromosomes, respectively.

1-2. Containing DNA: mitochondrion, nucleus, chloroplast; containing RNA: all.

1-3. Partially purified chromosomes contain large amounts of DNA; this can be detected by chemical tests and by spectrophotometry. Feulgen staining is a chemical test for deoxyribose, and during cellular mitosis, the stain is visible almost exclusively in chromosomes. When determined by microspectrophotometry of living cells during mitosis, the ultraviolet spectrum of a chromosome is similar to that of purified DNA.

1-4. One.

1-5. To estimate the quantity of cytoplasmic DNA, the number of mitochondria per cell must first be estimated.

1-7. 10^{-14} g $= 6 \times 10^{23} \times 10^{-14}$ or 6×10^{9} daltons. At 2×10^{6} daltons/μ, the length of a strand of DNA is $6 \times 10^{9}/(2 \times 10^{6}) = 3000\ \mu$. The volume of the DNA is about $0.3\pi\ (10^{-7})^{2} \approx 10^{-14}$ cm^3. The volume of *E. coli* (see 1-6) is $3 \times 10^{-4}\pi(1/2 \times 10^{-4})^{2} \approx 2.4 \times 10^{-12}$ cm^3; thus, $10^{-14}/2.4 \times 10^{-12} = 0.004$.

1-9. The bacteria are probably flagellated and highly motile, and hence, continually move away from the site of the parent bacterium of the colony. The methyl cellulose decreases the motility. The translucent film is a thin layer of bacteria which have moved from the main mass of the colony.

1-11. An anabolic reaction consumes energy; a catabolic reaction generates energy.

1-12. To generate ATP.

1-13. (b), (c), (d), (e), and (f) are true.

1-14. In ATP and other compounds containing high-energy phosphates, and in glycogen.

1-15. In the absence of O_2, the Krebs cycle is inactive, and less ATP is generated per mole of glucose.

1-17. Black produces only B gametes; gray makes both B and b, each with probability 1/2. Therefore, on the average, 1/2 of the offspring are BB and 1/2 are Bb. The probability of a single offspring being gray is 1/2. If there are two offspring, the probability of both being gray is 1/2 (for the first) $\times$ 1/2 (for the second) = 1/4. If there are three offspring, the probability of the first being gray and the others being black is $1/2 \times 1/2 \times 1/2 = 1/8$. Since the gray could be the second or the third offspring, the probability of having only one gray is 1/8 + 1/8 + 1/8 = 3/8.

1-18. One. The gametes are AB, Ab, aB, and ab. These mate to form 16 combinations, of which only $ab \times ab$ yields a homozygous recessive trait.

1-22. Recessive: (a),(c); dominant: (b),(d).

1-23. The gene order is $acbd$.

1-25. The three classes of frequencies represent single, double, and triple crossovers. If $ABCd$ and $abcD$ are the products of a single crossover, the order must be $ABCD$.

1-26. (a) In a cross $AbDe \times aBdE$, AE can result from a crossover anywhere between A and E. Since the map distance from A to E is $0.01 + 0.02 + 0.03 = 0.06$, AE will occur at a frequency of 0.06.
(b) To get $ABDE$, one needs crossovers in the intervals A–B, B–D, and D–E; the probability of getting all three cross-overs is $0.01 \times 0.02 \times 0.03 = 0.000006$. Therefore, the fraction of AE that will be $ABDE$ is $0.000006/0.06 = 0.0001$.

1-28. (a) Red and long are both homozygous recessive; white and short are both heterozygous.
(b) Homozygous dominant for either gene is lethal.

1-29. (a) Eye color.
(b) The eye color and wing length genes are both carried on the X chromosome of the archaeopteryx.

Chapter 2

2-1. As a simple repeating polymer it cannot carry information for an amino acid sequence.

2-3. (a) The solution containing 10^{-7} mg/ml DNA would contain 0.02 percent $\times 10^{-7} = 2 \times 10^{-11}$ mg $= 2 \times 10^{-14}$ g protein. The molecular weight of a protein containing 300 amino acids is 300 times the "average" weight of an amino acid, or, approximately $300 \times 100 = 3 \times 10^4$. We can set up the proportions

$$\frac{X \text{ molecules}}{2 \times 10^{-14} \text{ grams}} = \frac{6 \times 10^{23} \text{ molecules/mole}}{3 \times 10^4 \text{ grams/mole}}$$

which gives $X = 4 \times 10^5$ molecules.

(b) The *numbers* do not exclude the possibility that the transformation is protein-mediated, since the maximum estimate of the number of protein molecules (4×10^5) exceeds the number of transformants by 400-fold.

2-4. *Hint:* Measure the size of the molecules that can transform two genetic traits and compare the values to the sizes of those carrying only a single trait.

2-8. (b) After blending, most of the ^{32}P but little of the ^{35}S sediments with the cells.

(c) Some of the protein is injected into the bacterium.

(d) Some phages adsorb but fail to inject their DNA. Presumably they are defective.

(e) The ^{35}S-containing parts of the phages are removed from the bacterium, and phage production is still possible. Thus they are no longer needed for phage production.

2-9. *Hint:* Blending has apparently broken open the cells and the fragments of the cell wall plus any material that has adhered to them are in the pellet.

Chapter 3

3-1. The polar amino acids are arginine, asparagine, aspartic acid, glutamic acid, glutamine, histidine, lysine, serine, threonine, and tyrosine. The nonpolar amino acids are alanine, cysteine, glycine, isoleucine, leucine, methionine, phenylalanine, proline, tryptophan, and valine.

Isoleucine is more nonpolar than glycine, because it has a long, nonpolar side chain. Proline is unable to form a proper peptide because it lacks a free amino group.

3-2. Those with charged groups other than the α-amino carboxyl group bind metal ions. An ionizable protein can bind a metal ion simply by forming an ionic bond. Many metal ions (such as Mg^{2+} and Cu^{2+}) form complexes with amino groups. The Hg^{2+} ion binds SH groups and is frequently bound to methionine and cysteine.

3-3. Two cysteines cross-linked by a disulfide bond.

3-4. Tryptophan, tyrosine, phenylalanine.

3-5. All, using the CO and NH groups in the peptide bond. In addition, all polar amino acids can form hydrogen bonds via their side chains.

3-6. Electrophoretic mobility increases with charge but is reduced by friction between the moving molecule and the solvent molecules. Since valine has a longer side chain than alanine, it encounters greater friction and moves more slowly.

3-7. There are only five peptide bonds because the C—N bond between serine and proline is not a peptide bond.

3-8. (a) Since each molecule of protein X must contain at least one residue of the rarest amino acid, the number of amino acids in X cannot be less than $100/0.5 = 200$.
(b) The sedimentation rate is an approximate measure of the molecular weight, which, in turn, is an approximate measure of the total number of amino acids in a protein. A sedimentation rate slightly less than that of tryptophan synthetase A implies a number of amino acids less than 267. Evidently, protein X *does* contain 200 amino acids rather than some integral multiple of 200, and the number of methionines per molecule must be simply 3.

3-10. Once an enzyme has cleaved a terminal amino acid, it may then either cleave the next amino acid or remove the terminal amino acid from a second protein. Hence, when two amino acids per protein on the average have been removed, there are proteins which have had one, two, and three amino acids removed. Since only alanine has been found, the terminal tripeptide must consist of three alanines.

3-12. Think about the amino acids that would tend to form ionic bonds and hydrophobic bonds. The ionic bonds will be pH-sensitive.

3-15. Probably they are held together by an ionic bond.

3-17. Charged groups that normally are neutralized by ions present in the solvent interact and thereby change the shape of the protein from an active to an inactive form.

3-18. Foaming increases the surface-to-volume ratio of the solution. This increases the probability that an enzyme molecule will be at the surface and subjected to disruption by the force of surface tension. It is worth while to determine for yourself how these forces can destroy activity.

3-19. Three.

3-20. See 3-22 for reasoning.

3-22. The enzyme has three subunits if, in urea, the molecular weight drops three-fold. When active and inactive forms are mixed, dissociated, and reassociated, the subunits will reassociate *at random* to produce molecules having some active and some inactive subunits; those having 0, 1, 2, and 3 active subunits arise at frequencies 1/8, 3/8, 3/8, and 1/8, respectively. If the activities are 12.5 percent, 50 percent, and 87.5 percent, there must be 3, 2, and 1 active subunits, respectively, if the molecule is to be active.

Chapter 4

4-1. Pair each A, T, G, and C, with a T, A, C, and G, respectively.

4-2. *Bases:* adenine, thymine, guanine, cytosine, uracil; *ribonucleosides:* adenosine, guanosine, cytidine, uridine; *deoxyribonucleosides:* deoxyadenosine, thymidine, deoxyguanosine, deoxycytidine; *ribonucleotides:* adenylic acid, guanylic acid, cytidylic acid, uridylic acid; *deoxynucleotides:* deoxyadenylic acid, thymidylic acid, deoxyguanylic acid, deoxycytidylic acid.

4-3. Guanine-cytosine has 3.

4-4. CT, 0.15 (corresponding to AG since both are written in the $5' \rightarrow 3'$ direction and the DNA strands are antiparallel); AC, 0.03; GC, 0.08; AA, 0.10. If the strands were parallel, the known frequencies would be: TC, 0.15; CA, 0.03; CG, 0.08; AA, 0.10.

4-5. The mass/length of double-stranded DNA is approximately $2 \times 10^6/\mu$; hence, 32.8×10^6.

4-6. The order should be (a), (b), (c), since melting temperature increases with increasing GC content.

4-7. Molecule (a), since the long tract of GC pairs in (b) will require a higher temperature for strand separation.

4-9. Molecule (b), since the long tract of GC pairs will require a higher temperature for dissociation of intrastrand hydrogen bonds formed after denaturation and cooling.

4-10. Since stacking forces are a major factor in the stability of DNA, and since the terminal bases have nothing with which to stack, the terminal base pairs are frequently unbonded. The extent of loose bonding increases with increasing temperature.

4-12. The logarithm of the fraction of unbroken strands decreases linearly with time. Therefore, after 15 minutes the fraction unbroken is $[1 - (\frac{1}{3})]^3 = 8/27$. In general, when the fraction of unbroken strands is $1/e$, there is on the average one break per strand. From this fact, relative molecular weights can be calculated from the time yielding $1/e$ survival.

4-13. In A, because the temperature increases with the ratio

$$\frac{\text{Solubility of deoxyribose}}{\text{Solubility of bases}}.$$

4-14. (a) and (b).

4-16. (b), because the effective solubility of the bases would be increased.

4-19. (a) If the ionic strength is very low, the sugar-phosphate backbone of the separated strands and all regions thereof are highly charged and mutually repulsive, thus preventing reformation of base pairs. At high ionic strength both intra- and inter-molecular base pairs form and therefore OD_{260} decreases.
(b) When boiled, OD_{260} will increase to 1.37 but it will return to 1.00 at 25° C.
(c) After returning to 25° C some interstrand base pairs will form and OD_{260} will decrease somewhat.
(d) After returning to 25° C all base pairs will re-form and OD_{260} will decrease to 1.00.
(e) The ratio will be 1. The bases in each strand form a palindrome; that is, the sequence from bases 1 to 7 is complementary to the sequence from 14 to 8. Therefore each strand will fold back on itself and form a double-stranded hairpin-shape.

4-21. The mixture is 45/50 HH and 5/50 LL before denaturation. After renaturation there will be $(0.9)^2 = 0.81$ HH, $(0.1)^2 = 0.01$ LL, and $2(0.9)(0.1) = 0.18$ HL, or, 40.5, 9.0, and 0.5 μg of HH, HL, and LL, respectively.

4-23. Strand separation is incomplete in 100 percent methanol.

4-26. In 6 M sodium trifluoracetate, the melting temperature must be lowered so much that at 25° C the DNA is partly denatured—in fact, it must be 48.9 percent hyperchromic.

4-31. A linear DNA molecule can form a circle by cohesion of single-stranded termini; by recombination between redundant termini; and by exonucleolytic conversion of double-stranded termini to single-stranded termini, followed by hydrogen-bonding between single strands.

4-32. (b).

4-33. Force needed is $F_{1/4}$, since a circle has half the length and twice the width of a linear molecule having the same molecular weight.

4-35. Less ethidium bromide can be bound by a supercoil than by a linear molecule. Therefore, the decrease in the density of a supercoil is less than that of a linear molecule.

4-36. No, because ethidium bromide can intercalate between any two base pairs.

4-39. The supercoiled form, because it has many unpaired bases (that is, single-stranded regions) resulting from the unwinding of the helix.

4-40. The denser band is clearly a supercoil. When there is an average of one break per molecule, the fraction unbroken is $1/e$ or 37 percent. Therefore, the $2:1$ ratio would become $2(0.37):[1 + 2(1 - 0.37)]$, or, $0.74:26$, or, approximately $1:3$.

4-42. It is a circular DNA molecule consisting of one circular single strand (b), and one linear single strand (A) whose length is approximately 70 percent of the length of strand B. Thus the molecule is partly single-stranded, accounting for the results in (a), (b), (e), (f), and (g).

4-45. The DNA is double-stranded without single-stranded termini but with a single-strand break in each strand. These strand-breaks are each located $\frac{1}{3}$ of the distance from the 5′ end. The strand binding polyG has guanine at the 5′ terminus; the other strand has adenine. All mRNA is transcribed from a single strand.

4-47. A is a circle having a nick in each strand; B is a covalently closed circle, not necessarily supercoiled. C is a linear molecule. A and B have the same base sequence. X is a catenane consisting of one A linked with one B.

4-50. Treatment with phosphatase and polynucleotide kinase results in ^{32}P-labeling of the 5′-P termini. Annealing plus ligation produces a covalent circle. The ^{32}P is now attached to the base at the 3′-OH terminus. Enzyme digestion results in cleavage of the 3′-terminal base now labeled with ^{32}P. Therefore, guanine is at the 3′-OH terminus and, by complementarity, cytosine is the 13th base from the 5′-P termini.

4-51. Adenovirus DNA carries an inverted terminal repetition; that is, the sequence of bases along each strand is HO-3′-ABCD . . .D′C′B′A′-5′-P, where letters followed by primes denote the bases which can form base pairs with non-prime bases.

4-54. (a), no.
(b), no, because circles are not formed.
(c), left end is unique and constant, whereas the right end is variable.
(d)

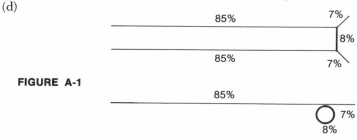

FIGURE A-1

Numbers = percent of length of entire DNA molecule

(e), in contrast with λ phage, the vegetative and prophage maps of mμ phage are colinear.

(f) (1), yes; (2), right end.

4-55. The DNA is about 30 percent unique, about 60 percent redundant, and the remaining 10 percent is highly redundant "satellite" DNA. The three classes have C_0t midpoints of about 10^3, 10^{-1} and 10^{-3}, respectively. Thus, the number of copies per genome is $10^3/10^3 = 1$ for the unique DNA (by assumption), $10^3/10^{-1} = 10^4$ for the redundant DNA, and $10^3/10^{-3} = 10^6$ for the satellite DNA.

4-56. Protein A may bind very weakly to superhelical DNA and may possibly introduce superhelical twists of the same sense as those in the DNA molecule. Alternatively, it binds strongly but does not alter s other than by increasing the molecular weight. Protein C may bind weakly and thereby remove some superhelical twists, but it may also be a highly asymmetric molecule that increases the frictional coefficient of the DNA molecule and thereby decreases s (by cancelling the effect of superhelicity). Protein B introduces superhelical twists in the opposite sense from those already in the DNA.

4-59.

FIGURE A-2

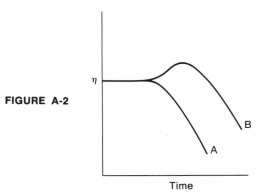

4-60. The irradiation probably damages the base so that the N-glycosidic bond (attaching the base to the deoxyribose) is broken by alkali. Once this bond is broken, the phosphodiester bond can also be broken by alkali. Thus the site of damage at each base is converted to a single-strand break. The slowly moving material consists of the DNA fragments formed by this breakage.

4-63. Single-stranded DNA can be distinguished from single-stranded RNA by enzymatic digestion by specific nucleases; RNA is hydrolyzed by alkali but DNA is not; the diphenylamine test distinguishes ribose from deoxyribose; acid hydrolysis followed by chromatography or electrophoresis distinguishes uracil from thymine.

4-64. Phage RNA contains ribosome-binding sites.

4-67. (a) There are three simple ways to identify the 3'-terminal fragment:

pre-charge the tRNA with a radioactive amino acid that will be attached to the 3'-fragment, use the "ACC enzyme" to remove pCpCpA from the 3'-end of tRNA and then replace the pCpCpA with radioactive nucleotides, or find the fragment that releases adenosine.

(b) The 5'-fragment is unique in that it will yield a diphosphoryl nucleotide (pXp) upon exhaustive degradation with alkali.

(c) Only that the 3'-terminal fragment contains 19 bases.

(d) Only that methyladenosine is on the 5'-end of the fragment.

(e) Fragment 1 is 3'-terminal because it has adenosine as a base hydrolysis product.

(f) Fragment 3 is 5'-terminal because it contains the methyl-A which is known from (d) to be 5'-terminal.

(g) Fragment 1: 5'-A; 3'-A.
Fragment 2: 5'-C; 3'-G.
Fragment 3: 5'-methyl-A; 3'-G.

(h) CpUpApGp.

(i) Current information is

5'-[methyl-A(2U,2C)Gp] [CpUpApGp] [A(2C,1A)pCpCpA]-3'

Fragment 3   Fragment 2   Fragment 1

The parentheses indicate that the composition but not the sequence of the segment is known. Assume a 3'-pCpCpA sequence with two G's distributed between fragments 3 and 2 or fragments 2 and 1. The best approach to finding the two missing G's is to digest the large (19-base) fragment with pancreatic RNase and find whether two GpGp's or one GpGpGp is generated, and what other nucleotides might be linked to it (or to them). The sequence of fragments 1 and 3 can be obtained by defining the products released by digestion with pancreatic RNase or with phosphatase and venom phosphodiesterase.

4-68. (a) Equimolar amounts of Ap, Cp, Gp, and Up indicate a tetranucleotide, since one Gp is the maximum possible after treatment with T_1 RNase. Therefore, the sequence at this stage is (A,C,U) Gp, in which the sequence of A, C, and U is not known.

(b) C must be 5'-terminal. Therefore, the sequence is either CpApUpGp or CpUpApGp.

(c) If the sequence were CpUpApGp, pancreatic RNase would yield free Cp and Up, which is not the case. Thus, the sequence is CuApUpGp.

The reaction scheme is now known to be

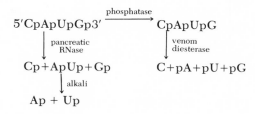

4-69. (a) pppGpCpUpUpCpCpCpGp.

(b) Mononucleotides.

(c) No, not at all. All internal nucleotides are pyrimidines, and all bonds would therefore be cleaved.

(d) pppGp is the initial nucleotide in transcript; tyrosyl tRNA is a cleavage product of 5s RNA.

4-70. G G G C T T A T C G A T A

4-71. (b) The sequence of the DNA in panel (i) is T C G A A G C C T T A A C.

4-72. The phage DNA strands are separated by denaturing, annealing to polyUG, and banding in a CsCl density gradient. Either light *or* heavy strands from each phage are annealed together; the *lac* DNA becomes double-stranded but the phage DNA stays single-stranded. The annealed DNA is digested with any DNase that attacks only single-stranded DNA. This treatment reduces the phage DNA to dialyzable nucleotides, whereas the *lac* DNA remains intact as the only pure species.

Chapter 5

5-1. (a) and (e) are true.

5-2. (c).

5-3. 3'OH.

5-4. No initiation: *dnaA⁻* and *dnaC⁻*; no synthesis: *dnaB⁻*, *dnaE⁻*, and *dnaG⁻*; no initiation of precursor fragments: *dnaG⁻*; reduced joining of precursor and uracil fragments: *lig⁻*, *polA⁻*; no DNA synthesis because of lack of deoxynucleotides: *dnaF⁻*; no effect: *polB⁻*.

TABLE A-1

	Fraction of DNA that is		
Generation	$^{15}N^{15}N$	$^{15}N^{14}N$	$^{14}N^{14}N$
0	1	0	0
$\frac{1}{2}$	$\frac{1}{3}$	$\frac{2}{3}$	0
1	0	1	0
$1\frac{1}{2}$	0	$\frac{2}{3}$	$\frac{1}{3}$

5-5. (a) If the DNA were not broken, the density would shift continuously; at one generation the DNA would have the density of $^{15}N^{14}N$ DNA.

5-7. X prevents initiation of a round of replication but has no effect on movement of the replication fork. Rifampicin or chloramphenicol are examples.

5-9. The two strands of DNA are antiparallel and the DNA polymers can only add a 5'-phosphoryl nucleotide triphosphate to a 3'-OH terminus.

5-10. Pulse-labeled DNA appears in small fragments when sedimented in alkali. If pulsed and chased, the radioactive DNA has the same average size as the major portion of the cellular DNA.

5-11. Measurement of the fraction of pulse-labeled material that is in small fragments. Hybridization does not give information because of bidirectional replication.

5-12. (a).

5-13. Polymerase I (polymerase), ligase.

5-14. (a); (d); (e).

5-17. The gap will be filled by a single piece. No, as long as one end of the gap had a 3'-OH group.

5-18. Bidirectionality of replication.

5-20. Add an inhibitor of DNA synthesis and see if pulse-labeled material is hydrolyzed. Alternatively, prepared DNA which is density-labeled at the origin—for example, by starving for an amino acid in the absence of label and then simultaneously restoring the amino acid and adding a density label. At a later time, place in a growth medium lacking the label and see if the label is lost. There are probably other experiments you may think of.

5-22. Since cytosine is deaminated to uracil, and uracil has different hydrogen-bonding properties, a mechanism must exist for removing uracil and replacing cytosine. A simple way to do this is to use an enzyme that removes all uracil. This method is not feasible if the uracil is base-paired with adenine in the DNA. Thus, DNA contains thymine and the enzyme that removes uracil does not recognize thymine.

5-25. In the replication of a circle the rotation of the replicating strands causes twisting of the nonreplicating portion. The ω protein cannot be involved in removing the twists because it can only remove twists in the opposite sense from those generated by replication. However, DNA gyrase can remove the twists.

5-26. This is a nearest-neighbor analysis that uses alkali instead of spleen diesterase and micrococcal nuclease (DNase II) to produce 3'-P mononucleotides from the RNA primer region and especially from the junction of RNA with DNA.
(a) DNA synthesized *in vitro* must have the general structure 5'-RNA-DNA-3'. At the RNA-DNA junction the sequence must be 5'-riboGp-deoxynucleotide.

(b) dATP is the first deoxynucleotide to be added onto the oligoribonucleotide primer.

(c) Some deoxynucleotide *other* than dA is added first to the oligoribonucleotide primer; the label remains in DNA, which is not hydrolyzed by alkali.

5-30. (a) All molecules in the act of synthesis will be completed but no resting molecule will be started. Therefore, a random population of cells will continue to synthesize DNA for one generation.

(b) There will be no synthesis since, during the period of amino acid starvation, all molecules in the act of synthesis will be completed.

5-31. DNA molecules in the act of synthesis will initiate a second round of DNA synthesis before the first round of replication is completed.

5-35. No, since it is unlikely to have an origin of replication.

5-38. No, since in both cases the replication forks have passed the point which is half the distance from the origin.

5-40. The resistance of circular DNA to exonucleases is one possibility.

5-44. The sex factor has probably integrated into the chromosome. Hence C contains both alleles. Yes, since *dnaA*⁻ and *dnaE*⁻ mutants are recessive.

5-45. The most recently synthesized DNA is probably membrane-bound.

5-47. Methylation of the bases to prevent self-destruction by restriction endonucleases.

Chapter 6

6-1. See Appendix B, "Properties of Selected Nucleases," page 238.

6-2. Almost all nucleases require the Mg^{2+} ion for enzymatic activity.

6-3. Alkaline phosphatase is used to remove terminal 5'-P, yielding 5'-OH. Polynucleotide kinase transfers ^{32}P from γ-^{32}P-ATP to the 5'-OH terminus.

6-4. The ligases form an active complex with AMP by reacting with either DPN or ATP.

6-5. To destroy foreign DNA.

6-6. Extracellular nucleases can break down nucleic acids present in the environment and resulting from cellular decomposition. The products of nucleolytic action can then be utilized metabolically. The pancreas secretes nucleases into the intestine to break down nucleic acids in the digestive process. Blood nucleases destroy nucleic acids released from viruses; this is essential since viral nucleic acid is frequently infective.

6-9. The enzyme fails to bind to the DNA at such high ionic strength.

6-10. The enzyme is inhibited by the nucleotides produced in the reaction.

6-11. The DNA molecule must contain something other than single-stranded DNA one-third of the way from the 5'-P terminus. Possibilities include protein or RNA linkages within the polynucleotide chain, a protein molecule bound to a sequence of DNA bases, a short sequence of double-stranded DNA or RNA, or a hairpin-like structure (a palindrome, or, inverted repetition). The boiling experiment indicates reversible thermal disruption at high but not low ionic strength; this strongly implicates the hairpin-like structure.

6-12. (b). A, pU, pC; Ap, UpC; Ap, UpC.

6-14. (a) A few single-stranded regions are produced at temperatures before which an increase in optical density is undetectable.
(b) Random breakage and re-formation of hydrogen-bonded regions occur continually. Because of the possibility of free rotation of the unbroken strand opposite a single-strand break, the fraction of time that the bases of any region are unpaired will increase in the vicinity of a single-strand break, thereby producing a single-stranded region. This will be more effective as temperature increases.
(c) Supercoiled DNA molecules possess single-stranded regions. A single strand break will first be made and then will become a double strand break according to the reasoning in part (b).
(d) A mismatched base pair generates a single-stranded region. Half of the renatured molecules will contain a mismatched base pair. Thus only the hybrid DNA molecules are attacked by S1 nuclease.
(e) The two phage strains probably differ by genetically undetectable single base changes. Thus, molecules of the type described in part (d) are produced.

6-15. Terminal transferase does not need a template. A poly dA strand can be added to one DNA molecule and a poly dT strand to another. The two molecules can be joined by hydrogen bonding between the poly dA and the poly dT.

6-17. Possibilities include a circular DNA molecule, altered termini (for example, altered 3'-P or 5'-OH groups), a blocked terminus (for example, blocked by an ester), and a "hairpin" producing a double-stranded terminus. The tests are left to you.

Chapter 7

7-1. Only bases 1 through 6 are shown.

(a) First round: One molecule each of

CATTAA

GTAATT

and

CTTTAA

GTAATT

Second round: Three molecules and one molecule, respectively, of

C A T T A A		C T T T A A
	and	
G T A A T T		G A A A T T

(b) First round: One molecule each of

(T T)		
C A A A		C A A A
	and	
G T T T		G T T T
		(A A)

Second round: Two molecules and two molecules, respectively, of

C A A A		C A T T A A
	and	
G T T T		G T A A T T

7-4. The ratio of polymerizing to exonuclease function will decrease in the antimutator since errors will be removed more often.

7-5. 5-Bromouracil is a base analogue substituting for thymine and sometimes pairing with guanine and inducing a transition. 2-Aminopurine substitutes for adenine and guanine and can pair with both thymine and cytosine and induces transitions. Nitrous acid causes deamination of guanine to xanthine, adenine to hypoxanthine, and cytosine to uracil. Hypoxanthine pairs with cytosine and uracil with adenine, so transitions are produced. Xanthine pairs with nothing. Acridine orange intercalates and produces frameshifts.

7-6. AAA, AGA, CAA, CGA, GAA, GGA, UUA, UCA, UAC, UAU, UGC, UGG, UGU.

7-7. If it is very near the terminus of the proteins.

7-8. The code is redundant.

7-9. The amino acid may not be in an active region of the protein. In addition, the exchange may not affect the conformation of the protein. Changes from polar to nonpolar amino acids and *vice versa* (from nonpolar to polar) are most likely to lead to mutation.

7-10. (a), (b), (c).

7-12. The change must be in the DNA strand which is transcribed, and it must be in a position within a codon that changes an amino acid.

7-13. Arg → His; Met → Ile; Gly → Asp.

7-14. In the first case, measure the reversion frequency; in the second, perform genetic crosses with known mutants and deletions and determine recombination frequencies.

7-16. (b).

7-19. A mutant allosteric enzyme whose regulatory activity is lost so that it is active when it should not be active, perhaps depleting a cell of an essential substance; a mutant protein that is part of a multi-subunit enzyme, for which a single defective subunit eliminates all activity; an operator mutation that binds a repressor so tightly that the repressor can not be removed; a mutation in an enzyme that have several activities (for example, *E. coli* polymerase I), which eliminates one activity and not the other, thus changing an essential ratio of activities. You may think of others.

7-20. A frameshift mutant should revert following treatment with acridines. A nonsense mutant should be reversed by putting it into a genetic background containing known suppressors. A missense mutant fails to respond to both of the preceding treatments.

7-21. Purple, purple, purple, purple, pink.

7-24. To select for a mutant which cannot synthesize X, grow bacteria in a medium lacking X and containing penicillin. Auxotrophs survive this treatment.

To select a temperature-sensitive amber suppressor, select a strain which is prototrophic for a substance Y by virtue of having an amber mutation in gene Y and possessing an amber suppressor, and grow at high temperature in a medium lacking Y and containing penicillin. A mutant with an amber suppressor inactive at high temperature will fail to suppress and will survive the penicillin treatment.

7-27. (a) $\log(1 - 0.95)/\log(1 - 0.05) = 60$.
    (b) $(0.95)(0.05)(100) \cong 5$.

7-29. Mutagenize the parent strain (for example, with nitrosoguanidine). Grow the mutagenized culture of the parent strain at 30° C in glucose-salts minimal medium containing G-3-P. Wash the cells in a salt solution, and resuspend in a medium that is similar but that contains ^{3}H-labeled G-3-P. Allow the cells to grow at 42° C for several generations. The wild-type cells will incorporate ^{3}H-G-3-P, and the desired mutants will not. Wash the cells in a salt solution to remove unutilized G-3-P and store at 5° C until ^{3}H-decay has killed 99.999 percent of the cells, as determined by plating at 30°. Test the surviving colonies for ability to grow at 42°. Check the *ts* mutants (that is, those which cannot grow at 42°) for ability to incorporate ^{3}H- or ^{32}P-G-3-P at 42° into material which sediments with bacteria and qualifies chemically as phospholipid. The desired mutants should not incorporate G-3-P into phospholipid.

Chapter 8

8-1. 5′CAUUGCAGCU3′ and 5′AGCUGCAAUG3′.

8-2. (b), (c).

8-3. (a, (b), (d).

8-4. Core enzyme: the β, β', α, ω tetramer; holoenzyme: core enzyme plus sigma factor; sigma factor: subunit responsible for recognition of the promoter; promoter: region of DNA containing RNA polymerase start sequences; operator: a repressor binding site.

8-5. This is the so-called "one gene–one enzyme" result; that is, a mutation in a particular gene never affects more than one enzymatic activity. In 1977 departures from this phenomenon were discovered in *E. coli* phage ϕX174.

8-7. The binding of RNA polymerase to DNA is sensitive to the number of superhelical twists.

8-9. Actinomycin intercalates and generally inhibits transcription. Rifampicin inhibits initiation of RNA synthesis. Streptolydigin inhibits chain elongation.

8-11. There is a mutation near the terminal region of cistron A which gives rise to a new and highly active promoter from which mRNA synthesis can occur under conditions of repression.

8-13. 5'-triphosphate and 3'-OH.

8-14. *Rho* factor.

8-15. (b), (c), (d).

8-17. (a) Yes, I, yes; (b) Yes, I, yes; (c) Yes, C, yes;
(d) Yes, I, yes; (e) Yes, C, yes; (f) Yes, I, yes;
(g) No, neither, no; (h) Yes, I, yes; (i) No, neither, no;
(j) Yes, no.

8-20. Messenger RNA is made shortly after addition of lactose and this is followed by enzyme synthesis. This continues for two generations, at which time depletion of lactose causes establishment of repression, turnoff of mRNA synthesis and hence of enzyme synthesis. Enzyme activity persists since β-galactosidase is fairly stable, although the activity per cell decreases as the cells divide.

8-22. (a), (b), (c).

8-23. (b), (c).

8-26. There might be a mutation in the gene for adenyl cyclase or for the cyclic AMP receptor protein. Another possibility is a transport mutant.

8-27. (c).

8-28. Two. One if glucose is present.

8-29. When the *lac* operon enters the female, no *lac* repressor is present so that z-y transcription occurs and β-galactosidase is made. However, shortly after transfer, repressor mRNA, and hence, repressor, are made.

This will shut off z-y transcription, and hence, will shut off enzyme synthesis. If the female is $i^+ o^c z^- y^-$, repressor is present and no enzyme is made. The o^c in the female is irrelevant.

8-33. There are frequently several copies of an F$'$ sex factor per cell and each can be transcribed and produce enzyme.

8-34. (a)   **TABLE A-2**

	Expected results	
Experiment	Induced	Uninduced
5	100	0.1
6	100	0.1
7	100	100
8	100	100
9	0.1	0.1
10	0.1	0.1

(b) There is twice the amount of β-galactosidase activity because there are two z genes, both making enzyme.

(c) The most probable reason is that the z_1^- mutant protein is reducing the activity of the good z protein by **negative complementation.** In positive complementation, two defective proteins complement each other to give a good, or semigood protein. In negative complementation, a defective protein complements with a good protein in a negative manner; that is, a defective one makes a good protein bad. A careful examination of experiments 5 and 8 should convince you that only when z_1^- is induced is the activity of z^+ reduced.

Negative complementation can occur because β-galactosidase is a tetramer consisting of four identical polypeptide chains. In a diploid there is free mixing of the subunits made from each *lac* region. If one defective z^- peptide chain comes together with three other good z^+ peptide chains, it "poisons" the good chains and reduces their activity.

(d) The explanation is **positive intragenic complementation;** when defective subunits z_2^- and z_3^- mix to form a tetramer, they make each other "good."

8-35. At very low levels all inducible operons will be transcribed—for example, a repressor might come off the operator for an instant and RNA polymerase will get on. Hence each cell will contain a few β-galactosidase and permease molecules. The permease will let in a few lactose molecules and these will be converted to the true inducer, *allo*-lactose; then derepression can occur.

8-36. RNA is hydrolyzed to short oligonucleotides in alkali. These do not sediment. If the fractions had been treated with acid and filtered to collect acid-insoluble radio-activity, no material would have been found at the meniscus and the worker would have realized that something was wrong.

8-39. Site *a* is an operator; *b* is the structural gene; and *c* is a regulator.

8-42. A deletion removing the region *z* through *tsx* has resulted in fusion of the *y* gene to the *pur* operon. Transcription of the *y* gene no longer proceeds from the *lac* promoter, but from a promoter in the *pur* operon that is regulated by the *pur* operator.

8-44. (a) The protein *phoR,* because in its absence the expression is insensitive to the presence or absence of phosphate.
(b) The protein *phoB;* it is a positive effector. The *phoB* protein must be binding to DNA and acting as a positive effector necessary for synthesis of APase. If the *phoR* protein were acting directly on the DNA, then it would have to be a repressor. Mutants of the *phoR* gene do show constitutivity, as repressor mutants do, but a *phoB⁻ phoR⁻* double mutant would also show constitutivity if the *phoR* gene were a direct repressor, and they do not. Therefore, since the *phoR* protein cannot be a direct repressor of transcription, the *phoB* protein must be a direct positive control element affecting transcription of the *phoA* gene.
(c) If the *phoB* protein binds to the DNA at a promoter-like spot, and the *phoR* protein in the presence of phosphate acts to reduce the synthesis of the *phoA* protein, then the *phoR* protein probably forms a complex with the *phoB* protein and phosphate, making it harder for the *phoB* protein to bind at the promoter site. Thus, transcription is inhibited. The *phoB⁻* mutation allows only 5 percent expression of the *phoA* gene in the absence of phosphate. In the presence of a *phoR⁺* allele the residual expression is still sensitive to phosphate expression (the expression is reduced to 1 percent) but in the presence of the *phoR⁻* allele the 5 percent expression of APase cannot be reduced by phosphate.

8-47. (a) First, repressor cannot bind X; the operon is on. Second, repressor cannot bind to operator even when X is present; the operon is on. Third, repressor binds operator without X; the operon is off.
(b) Phenotypes of partial diploid with wild-type and each of the mutants are: (first) inducible; (second) inducible; (third) unable to synthesize X under all conditions.

8-50. (a) lambda *bio* H, yes; lambda *bio* L, yes; lambda H, no; Lambda L, no. Transcription is divergent (goes in both directions).
(b) A, unchanged; B, unchanged; F, decreased; C, decreased; D, decreased. The polar mutant affects only genes downstream from the promoter.
(c) It is a *trans*-dominant superrepressor mutant which does not undergo any allosteric change in the absence of biotin and therefore does not come off the DNA at the operator site when biotin is present.

(d) The bacterium seems likely to contain a naturally occurring mutation of the type described in part (c).

8-52. The only difference between the transcription maps of T3 and T7 DNA
is at the end of the early region. With T7 DNA, RNA polymerase
terminates without requiring the *rho* factor; with T3 DNA the *rho*
factor is needed.

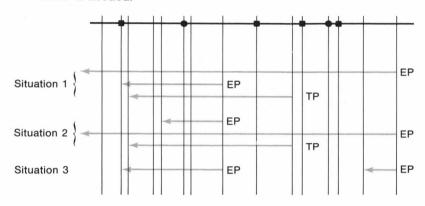

(b)   **TABLE A-3**

Situation	Strength of expression			
	B	D	F	G
1	1	2	3	3
2	1	2	2	2
3	0	0	1	1

8-56. The region from 45 to 60.

Chapter 9

9-1. Met, Pro, Leu, Ile, Ser, Ala, Ser.

9-2. (a), (c), (d).

9-3. Histidine, glutamine, cysteine, tryptophan, serine, glycine, leucine,
proline, isoleucine, threonine, asparagine, lysine, and methionine.

9-4. For UAG, the amino acids are Tyr, Leu, Trp, Ser, Lys, Glu, and Gln.

9-6. Arg-2 could be replaced by Gly, Trp, Lys, Thr, or Ser. Arg-3 could be
replaced by Ser, Lys, Thr, and Gly.

9-9. Val-Cys-Val-Cys-Val-Cys. . .

9-11. In any sequence of three bases in which A occurs five times as often as C, the probabilities of having A or C in a particular position are 5/6 and 1/6, respectively. Therefore the probabilities for each triplet are: AAA, $(5/6)^3 = 125/216$; ACA, AAC, CAA, $5/6 \times 5/6 \times 1/6 = 25/216$ for each; ACC, CAC, CCA, $5/6 \times 1/6 \times 1/6 = 5/216$ each; CCC, $(1/6)^3 = 1/216$. If the relative amount of AAA is defined to be 100, then the relative amounts of each of the other triplets is: ACA, AAC, CAA, $25/125 \times 100 = 20$ each; ACC, CAC, CCA, $5/125 \times 100 = 4$ each; $1/125 \times 100 = 0.8$. Therefore, the codon assignments are: Lys, AAA; Asn, triplet having 2A and 1C; Gln, triplet having 2A and 1C; His, triplet having 1A and 2C; Pro, CCC and a second triplet having 1A and 2C; Thr, two triplets, one having 2A and 1C and the other, 1A and 2C.

9-12. (a) Arg, Asp, Thr.
(b) Met, Asp.
(c) Ile, Asn.

9-13. (a) If A were added, the terminating codons would be UUA, UAA, and AAA, in decreasing frequency. The amino acids attached to polyphenylalanine would be leucine and lysine (UAA is a stop codon). Results for C and G are similarly obtained.
(b) No. Thirty-six triplets would not be determined.

9-16. AUG might have to be near a ribosome binding site or some other special sequence in order to serve as a start signal. A stop codon will not cause termination when it is not in the reading frame.

9-17. The Tyr and His codons followed by a terminator UAA codon are easy to find in the mRNA. To add two more amino acids, one has to add a nucleotide to make two extra codons. This must be done in such a way as to break up the terminator codon, adding one base and generating codons for Leu and Ser. AGC codes for Ser, so something must be added to the U and A of the termination codon to get a Leu codon. The only way to do this is to add a U on either side of the U in the termination codon. Thus there must be a frameshift mutation adding U to give the sequence 5'-pAAGUAUCACUUAAGC.

9-18. (a) The transcript from DNA strand 2 is

5' -X-AUA-UAG-GGG-GCA-Y- 3'.

The transcript from DNA strand 1 is

5' -Y'-UGU-CCC-CUA-UAU-X'- 3'.

The DNA strand 2 transcript contains a nonsense codon (amber, indicated with an underline) so that the coding strand is 1 and the amino acid sequence is NH_2-Cys-Pro-Leu-Tyr-COOH.
(b) A single-step mutation of the UGC codon yields UGA, a nonsense codon; mutation of the UAU codon yields UAA or UAG, both nonsense codons.

(c) The mRNA sequence for such an amino acid sequence is

$$UG \begin{pmatrix} U \\ C \end{pmatrix} CC \begin{pmatrix} U \\ U \\ C \end{pmatrix} UA \begin{pmatrix} U \\ C \end{pmatrix} AUG,$$

where the symbols in parentheses represent alternate bases, each of which would yield the same amino acid sequence. Given the base sequence of the wild-type, the mutant sequence must be Y'-UGC-CCC-UAU-AUG, which cannot be generated from the original sequence by a single base substitution. Instead, a deletion of a C in position 3, 4, 5, 6, or 7 (starting with UGC) must have occurred.
(d) From (c), and since the third base of the Met codon is G, X' must be G and X must be C.

9-19. (a) Alanine, $GC \begin{pmatrix} U \\ C \end{pmatrix}$, is derived from asparagine, $AA \begin{pmatrix} U \\ C \end{pmatrix}$, by two base changes; that is, $G \rightarrow A$ and $C \rightarrow A$.

(b) Asparagine, $AA \begin{pmatrix} U \\ C \end{pmatrix}$, is converted to serine, $AG \begin{pmatrix} U \\ C \end{pmatrix}$, by a transition. It is converted to lysine, $AA \begin{pmatrix} U \\ G \end{pmatrix}$, by a transversion.

(c) Threonine (ACG) $\rightarrow$ methionine (AUG); that is, $C \rightarrow U$.

9-20. The lysine and glycine codons must be $AA \begin{pmatrix} A \\ G \end{pmatrix}$ and $GG \begin{pmatrix} A \\ G \end{pmatrix}$, respectively, to have been derived from arginine by a single base substitution. Recombination between positions 1 and 2 in these codons yields codons $AG \begin{pmatrix} A \\ G \end{pmatrix}$ and $GA \begin{pmatrix} A \\ G \end{pmatrix}$. The first code the original amino acid arginine and the second code glutamine.

Chapter 10

10-1. (a) Met, His, Tyr, Cys, Ser, Gly, Ala.
(b) Met, His.
(c) UAU, UAC, UUA, UCA, AAA, GAA, CAA.

10-2. (c), (d).

10-3. (a).

10-4. (a), (b), (d).

10-5. (a), (b), (c).

10-6. (b).

10-7. (a).

10-8. Only an approximate calculation can be made since all tRNA genes and the number of copies of each are not known accurately. A tRNA has a molecular weight of approximately 25,000. This is copied from double-stranded DNA so the DNA would have a molecular weight of 50,000. We might assume there are no more than 61 tRNA's (one for each codon), but if wobble is considered, 40 is probably more reasonable. In most cases examined carefully there are 2 copies of each tRNA gene, so that the amount of DNA is $5 \times 10^4 \times 40 \times 2 = 4 \times 10^6$ daltons. In some cases, we know that a functional tRNA molecule is cut down from a molecule roughly twice as large. Hence 8×10^6 daltons is probably a reasonable value. This is $(8 \times 10^6)/(2.6 \times 10^9) \approx 0.3$ percent of the *E. coli* chromosome.

10-9. The average molecular weight of an amino acid is 100. Thus there are 500 amino acids and 3,000 base pairs corresponding to this. The molecular weight of the DNA is 10^6.

10-10. Less, because with polyribosomes several protein molecules can be synthesized simultaneously from a single mRNA molecule.

10-11. (b).

10-12. (a). Be sure to start at the AUG codon.
(b) No; it is not in the reading-frame generated by the AUG codon.

10-13. Free carboxyl group.

10-14. (a) Met Lys Lys Lys. . . .
(b) Met Lys.

10-18. (a) Mutation generates sites subject to *rho*-dependent termination of transcription. This does not always occur, so that some mRNA molecules are made normally.
(b) There are several ribosomal binding sites on the mRNA and they have different affinities for the ribosome. Alternately, the AUG start codons for each cistron do not all have the same start-efficiency, depending on adjacent sequences.

10-19. Release of finished proteins from the ribosome.

10-22. Synthesis would have to await the completion of a molecule of mRNA. In the existing system protein synthesis can occur while the mRNA is being copied from the DNA. Thus, protein synthesis can start earlier than would be possible with reverse polarity, and the mRNA is relatively resistant to nuclease attack.

10-23. (a): (3), Label is found at -COOH end only.
(b): (2), Label in amino acids at all positions.

10-24. Formylmethionyl-tRNA and mRNA plus (perhaps) other things are required *in vivo;* a high Mg^{2+} concentration is required *in vitro.*

Chapter 11

11-1. The original mutation has produced an amino acid change that affects an interaction with some other part of the protein. This alters the shape of the protein. A change of an amino acid in the second part of the molecule can restore the interaction. A simple example would be when amino acid A, which is positively charged, binds to B, which is negatively charged, simply by charge attraction. In the mutant, A is replaced by C, which is negative. A change of B to D, which is positive, would restore the *status quo*. This is an example of "two wrongs making a right."

11-2. A missense suppressor replaces one amino acid by another. It can be a mutant aminoacyl synthetase. A nonsense suppressor inserts an amino acid at the site of a chain-termination codon.

11-4. There are two possibilities: (1) There are two genes for the tRNA species that is mutated to yield the nonsense suppressor, and only one of these has been mutated. (2) Natural chain-termination sequences might usually consist of two or more different termination codons. The nonsense suppressor will suppress only one and chain termination will still occur. Both (1) and (2) are probably important.

11-7. A tRNA molecule could have a four-base anticodon.

11-8. Some of the possibilities are (1) a mutant aminoacyl synthetase that recognizes the wrong tRNA, (2) a tRNA mutated in the recognition loop so that it is sometimes charged with the wrong amino acid, (3) a tRNA mutated in the anticodon, and (4) a tRNA mutated in the anticodon loop next to the wobble base in such a way that the codon–anticodon specificity is reduced. Missense suppressors must be weak in order that all normal proteins do not become alterers. A Y-for-X missense suppressor need not be active against all X-type mutants because substitution of Y may not always result in a functional protein.

11-10. A tRNA molecule having a four-base anticodon could suppress a base addition. Such a mutant tRNA molecule could be produced by acridine mutagenesis since the acridines produce base additions. A single base deletion would have to be suppressed by a tRNA molecule having a two-base anticodon. This is not impossible but seems unlikely because, if it occurs, many of the normal tRNA molecules having three-base anticodons would be able to suppress the deletion and therefore the deletion would not be picked up as a mutant. On the other hand, perhaps such suppression does occur, but because suppression is weak and because so many different amino acids substitutions would occur at the mutated site, the number of active protein molecules is so low that reversion is not detectable.

11-11. The G to C mutation creates a UAG "stop" signal. Thus the AUG found "downstream" (and in the proper reading frame) can start in-frame synthesis going again.

11-13. The mutation *dnaB*266 is a suppressed amber mutation. The *supD* and *supE* suppressors insert amino acids which cause the protein specified by the *dnaB* gene to be temperature-sensitive. The *supF* suppressor inserts an amino acid which causes the *dnaB* protein to be stable at high temperature. In the absence of a suppressor, the complete *dnaB* protein is not made, and the strain having the *dnaB⁻ sup⁻* genotype is therefore nonviable.

11-16. (c)

Chapter 12

12-1. Activity can be regulated by dissociation and association of subunits. For example, in a protein solution at low concentration, an inactive monomer will be the major species.

12-2. Enzymes whose activity is regulated by the presence of small molecules (for example, allosteric enzymes) must contain several binding sites and therefore tend to be large. The polymerases that carry out several functions (for example, polymerase, $5' \rightarrow 3'$ exonuclease, and $3' \rightarrow 5'$ exonuclease) must have numerous active sites. A typical protein with a single active site usually has a molecular weight of 20,000–60,000. With active sites for binding of bases in polymerization, plus the other activities, a molecular weight in the 100,000–200,000 range is expected.

12-4. Metal ions or small molecules are part of the active enzyme; these are the electron acceptors.

12-6. (a) G inhibits 5; H inhibits 7; G and H together or E alone probably inhibit 3; J inhibits 8; enzyme 1 could be inhibited by (G, H, and J), (E and J), (G, H, and C), or (C and E); the ( ) enclose the products that act together.
(b) The steps inhibited by several products may use isozymes.

12-8. A competitive inhibitor can bind to an active site of an allosteric enzyme and increase the substrate-binding affinity of other sites on the same molecule.

12-9. (a), (1) Can be highly cooperative depending on L_0; (2) Not cooperative; (3) Not cooperative, no T state, saturation of R state only; (4) Not cooperative, no R state, saturation of T state only; (5) Highly cooperative; (6) Not cooperative.
(b) The ratios are calculated from $L_n = L_0 c^n$. Hence $L_1 = 10^2$, $L_2 = 1$, $L_3 = 10^{-2}$, and $L_4 = 10^{-4}$.

12-10. To compare the monomer and the tetramer for amplification properties, draw graphs for saturation of each, using values of $\bar{y}$ generated from the Monod-Wyman-Changeux equation for various values of α.

For the monomer: $\bar{y} = \dfrac{\alpha}{11 + \alpha}$.

For the tetramer: $\bar{y} = \dfrac{\alpha(1 + \alpha)^3}{10^4 + (1 + \alpha)^4}$.

From the graph, determine the values of α corresponding to $\bar{y} = 0.1$ and $\bar{y} = 0.9$ for the two cases; these are 1.22 and 99 for the monomer. Since α is directly proportional to the concentration of F, there is a $(99/1.22) = 81$-fold change in the concentration of F required to change $\bar{y}$ from 0.1 to 0.9.

12-11. (c) The mutant enzyme has $\frac{1}{4}$ the molecular weight of the wild-type, so is probably a single-subunit enzyme (it cannot associate to form tetramers). Therefore, there is no poisoning effect for two reasons: (1) The mutant enzyme subunits are still active, so we would not expect to find reduced activity in a mixture of mutant and wild-type subunits. (2) The mutant subunits cannot associate to form tetramers with each other, so we would not expect them to be able to form hybrid enzymes with the wild-type subunits. Therefore, even if the mutant subunits had greatly reduced activity, we would not expect them to be able to "poison" the activity of the good subunits.

12-12. (a) and (b) The molarity of cyclic AMP (cAMP) inside the bag is the molarity of unbound cAMP plus the molarity of bound cAMP. The molarity of unbound cAMP inside bag equals the molarity of cAMP outside bag. Call this [A]. Therefore the molarity of bound cAMP can be determined by subtraction, and from the molecular weight of the protein and the concentration of protein in the bag, one can calculate N, the number of molecules of cAMP bound to each molecule of protein. One can then make a Scatchard plot, that is, $N/[A]$ versus N. The y-intercept yields n/K_D; the x-intercept is n. Therefore $K_D = 10^{-5}$ M and $n = 1.6$.

(c) There are several possibilities. The molecular weight estimate of 6×10^4 may be incorrect (this value was obtained by sucrose-gradient zone-centrifugation, which is not highly accurate). The protein preparation may have been impure and contained degraded cAMP-binding protein which was unable to bind cAMP, or it may have contained some other protein. Some irreversible inhibitor of cAMP binding might have masked some sites.

12-13. Pseudo-first-order kinetics, like first-order kinetics, are described by the equation

$$\log_{10} \frac{A}{A_0} = -\frac{kt}{2.3} \ ,$$

where A is the enzyme activity at time t, A_0 is the starting enzyme activity, and k is the rate constant for the reaction. The values given in the problem can be plotted on semi-log paper to yield $k = 0.154$ min^{-1}. Thus for 99.9 percent inactivation, $t = 45$ min.

12-14. The two main possibilities to explain temperature-sensitivity are (1) shifts in equilibrium and (2) change in inactivation rate.

Case 1: If 90 percent of the enzyme is *active at* 30° C and only 10 percent is *active at* 42° C, the protein-folding reaction would have to have a very large positive $\Delta H°$, roughly 70 kcal, and a large $T\Delta S$ term of about 70 kcal to balance ΔH in the 30°-to-42° temperature range, so that $\Delta G°$ would be zero near 36° C. The $\Delta G°$ of the wild-type protein would be more negative, indicating its stability in the range from 30° to 42°, but the wild-type protein would denature sharply at a higher-temperature range, say 50°–60°. Protein denaturation is known to have a high $\Delta H°$, that is, to be highly sensitive to temperature.

Case 2: For a great temperature dependence, where the rate of inactivation is slow at 30° C and fast at 42°, the reaction must have a large positive ΔH. For example, an inactivation rate would increase 100-fold from 30° to 42°, if $\Delta H = 70$ kcal, as can be calculated from the two equations

$$\log \frac{k_2}{k_1} = \frac{E_a}{2.3R} \left(\frac{1}{T_1} - \frac{1}{T_2}\right) \quad \text{and} \quad \Delta H = E_a - RT,$$

in which k_1 and k_2 are the rate constants at temperatures T_1 and T_2, E_a is the activation energy and R is the gas constant. Also for the half-life of the enzyme to be short at 42°, say, 17 seconds, ΔG must be about 20 kcal, so the $T\Delta S$ term must be large, for example, 50 kcal, to offset the large ΔH. Then at 30° the lifetime of the enzyme will be roughly 30 min.

12-15. Start with the equation

$$K_D = \frac{[\alpha][\beta]}{[\alpha\beta]} \ .$$

One hundred molecules per 1 $\mu^3 = 10^{17}$ per liter or 1.7×10^{-7} M. Since 90 percent α and β are in a complex, the molarity of α (and β, since $[\alpha]$) is 1.7×10^{-8}. Therefore,

$$K_D = \frac{(1.7 \times 10^{-8})(1.7 \times 10^{-8})}{(0.9)(1.7 \times 10^{-7})} = 1.9 \times 10^{-9}\,M.$$

12-17. Analyze behavior of β-galactosidase mutants in terms of the lysozyme mechanism, the main point of which is that the maximum binding energy available at the lysozyme active site is partially diminished by the amount of energy required for the substrate to reach equilibrium after conversion from chair- to half-chair form, so that the *observed* binding energy is less than the maximum. However, the stabilization of the half-chair form of the substrate on the enzyme surface brings the ES complex closer to the transition state for the substrate and effectively lessens ΔG. Thus, binding energy is used to accelerate the reaction.

Mutant 1: V_{max} is low and K_D is small (that is, the enzyme-substrate affinity is *high*). Possibly in this galactosidase mutant, K_D is smaller because the maximum binding energy of the site is no longer offset by the strain energy needed for formation of the half-chair substrate. That is, the mutant enzyme binds the nonstrained substrate preferentially.

V_{max} is lowered because the nonstrained substrate is farther from the transition state. In short, most ES complexes are nonproductive.

Mutant 2: V_{max} is increased and K_D is higher (that is, affinity is *lower*). Higher V_{max} may indicate that substrate is still more like the transition state, on the average, when bound. That is, the mutant enzyme more effectively selects the half-chair form of the substrate from the substrate population and does not occupy its site as often with the non-half-chair configurations of the substrate. The ES complexes would comprise fewer "nonproductive" complexes than are found for the wild-type. For example, in lysozyme, the hexasaccharide is thought to bind to the active site in the nonproductive mode in about half of the ES complexes (such as binding to sites A—B—C with the three rings at the aldehyde end of the hexasaccharide—binding as if a trisaccharide). Mutant β-galactosidase would reduce the frequency of nonproductive complexes, perhaps by binding less well at the galactose end of lactose. However, K_D is increased overall since now the total binding energy of the site is less and is offset by the same large, unfavorable strain energy.

Mutant 3: V_{max} is greatly reduced and K_D is increased (that is, there is less affinity). Lactose binds less well and when it does, less frequently, in the transition-state configuration. Possibly a region of the site which bonds with lactose in both its unstrained (unreactive) and strained (reactive) configurations is damaged, and the damage is such that the strained form can no longer line up with the remainder of the site, at which additional bonding would compensate for the strain energy. There are other equally plausible explanations.

Chapter 13

13-1. (a)

13-3. (b) is true for both (1) and (2).

13-4. (a) X is not inducible.

(b) X would be considered inducible. The residual 5 percent could be due to either a second noninducible system which can excise thymine dimers or to leakage of synthesis of X proteins past the chloramphenicol block.

13-7. B/r probably existed in the population before irradiation and was recovered since it had a higher probability of survival than the wild-type. This could not have been the case for B_s, which would not have survived. Therefore it seems reasonable to think that B_s arose from a mutation induced by the UV-irradiation. If this were so, the colony would be expected to contain wild-type cells, since parental cells are always produced in the first division after mutagenesis. However, this also seems unlikely, since the original screening test for B_s consisted of showing that the colony contained mostly sensitive cells; if 50 percent were resistant, Hill would not have recognized the colony as one containing mostly sensitive cells. Therefore, it seems likely that the parental DNA strand—that is, the strand containing the wild-type base sequence for the B_s gene—must have suffered additional damage (for example, a lethal mutation), so that it would not give rise to viable progeny. Other explanations can probably also be given.

13-10. If DNA replication occurs before sufficient repair of the damage is done, nonfunctional DNA might be formed.

Chapter 14

14-1. The DNA is coated with molecules of cytochrome C. This increases the effective thickness of the DNA. Shadowing at a low angle allows large amounts of metal to accumulate against the thickened molecule in the manner that snow accumulates against a fence in a storm.

14-2. In the negative contrast technique, the biological sample (which is transparent to electrons) is surrounded by an opaque material—in this case, tungsten atoms. The intensity of the image is inversely proportional to the thickness of the opaque material. Hence, a phage head filled with DNA appears as a solid white object against a darker background, but an empty phage head will become filled with phosphotungstic acid and will therefore appear as a light hexagonal outline against a dark background.

14-3. (a) If θ is the shadow angle and l is the length of the shadow produced by a particle of height L, then $L = l \tan \theta$. The shadow angle can be accurately determined from the length of shadow of a second particle of known dimensions.
(b) 492 Å. The fuzzy ones are probably partially collapsed molecules.

14-4. Such metals would adsorb more electrons than the sample; hence, contrast would be diminished.

14-6. (a) Protein binds to nitrocellulose, and DNA does not. Therefore, a DNA molecule to which protein molecules are bound will adhere to the filter by means of the protein molecules.

14-7. (a) Gel chromatography or gel electrophoresis.
(b) Add proteins of known molecular weight and obtain a standard curve in which either the volume of the effluent (for chromatography) or the distance migrated (for electrophoresis) is plotted against molecular weight.

14-8. Hydroxyapatite chromatography; filtration through nitrocellulose filters; digestion with nucleases specific for single-stranded DNA.

14-12. No. To move an object x units north and then y units east brings the object to the same place reached by moving it first y units east and then x units north.

14-13. (a) No, since differences in shape can affect mobility—that is, two proteins differing in both molecular weight and shape might have the same electrophoretic mobility.
(b) Yes.

14-17. Most restriction endonucleases generate DNA fragments having complementary single-stranded termini which interact to form short double-stranded segments capable of holding two or more fragments together. These double-stranded segments dissociate at 65° C.

14-21. Three tenths are dimers.

14-22. Thirty-two percent and 57 percent of the molecules are broken by the first and second doses, respectively. To calculate the rate, one needs to calculate from the graph the dose D_{37} yielding $1/e$ or 37 percent surviving strands; this is the dose at which on the average there is one break per strand. This is 5,000 rads. Thus the rate of breakage is

$$\frac{1 \text{ break}/5{,}000 \text{ rads}}{20 \times 10^6 \text{ daltons}},$$

which is commonly expressed as

$$\frac{10^{-5} \text{ breaks/rad}}{10^6 \text{ daltons}}$$

14-23. (a) The molecule is compact and probably is roughly spherical.
(b) The molecule is an extended rod.

14-24. $(1.7 + 1.3)/2 = 1.5 \text{ g/cm}^3$.

14-25. To work this problem, you must know the average nitrogen content and carbon content of a protein, and you must assume that isotopic substitution does not change the volume of the protein (a good assumption).

14-27. The density of the virus particle is less than that of $3\ M$ potassium tartrate. This low density probably results because the virus contains lipid molecules or lipoprotein molecules. When these are removed by ether, the density increases.

14-28. A flexible rod encounters less friction when moving through a fluid.

14-29. The molarity of the solution is $(1/423)0.032$. The molar extinction coefficient is $0.27/(0.032/423) = 3{,}569$.

14-30. Make a graph of relative concentration versus OD. From the linear region in which Beer's Law is obeyed, the OD the original solution would have if Beer's Law were obeyed can be calculated. This is 4.20. Thus the molarity of $4.20/348 = 0.123$.

14-32.

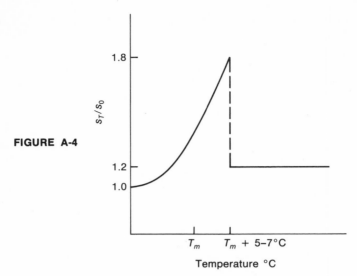

FIGURE A-4

Chapter 15

15-1. By destruction of host DNA; by conversion of host RNA polymerase to a form which cannot read host promoters; by inactivation of host RNA polymerase and replacement with phage-coded polymerase.

15-2. T4 will be turbid. T4h will be clear.

15-3. (a) Smaller phages diffuse further in agar.
(b) If a phage is adsorbed to a bacterium before dilution into agar, new phages will be produced sooner than if the phage must search for a bacterium in the agar. Since there is only a limited time, in agar, in which phages can be synthesized (until bacterial growth stops), the size of the plaque depends upon how quickly a phage particle adsorbs to a bacterium.
(c) The halo is caused by lysis of some of the bacteria surrounding the plaque, the result of lysozyme which is released by cells within the plaque and which diffuses outward.

15-6. Use Poisson's law,

$$P(r) = \frac{x^r e^{-x}}{r!},$$

where $P(r)$ = fraction of cells adsorbing r phage, x = moi, and r = the number of phage adsorbed to an individual cell. The fraction of cells infected by two or more phage = $1 - P(0) - P(1) = 0.8$. Therefore, 80 percent of 10^8 bacteria (= 8×10^7 bacteria) are infected by two or more phage.

15-8. From Poisson's law, e^{-2} or 0.137 of the bacteria are uninfected. Therefore the number of plaques is $200 (1 - .137) = 173$.

15-11. (T3, T7); T1; T5; (T2, T4, T6); parentheses group T phage species having DNA with the same molecular weights.

15-12. The head.

15-13. After adsorption, divide the mixture in half. Add antiserum to one half to kill unadsorbed phages. Measure number of infected cells by their ability to produce phage. This is the number of infective centers. To the other half, add chloroform to kill infected cells and measure unadsorbed phage. The multiplicity of infection equals

(Phage added − Unadsorbed phage)/Infected cells.

15-14. The original lysate contains nonviable particles which can complement the defect in ϕp^- but not in ϕq^-.

15-15. *Infectious center:* a bacterium capable of releasing phage; *multiplicity reactivation:* the production of viable phage by a bacterium infected with a high multiplicity of nonviable phage; *eclipse phase:* a stage in the life cycle in which no infective phage particles can be found; *marker rescue:* the appearance of a wild-type allele in a viable phage from a bacterium which is infected with a nonviable wild-type phage and a viable mutant; *lysis-from-without:* the lysis of a bacterium by a large number of phages (which may all be noninfective) and without the phages entering the normal life cycle; *heterogenote:* a transductant containing both the original mutant allele from the recipient and the wild-type allele from the donor; *phenotypic mixing:* in a mixed infection with two phages having different protein coats, the production of phages with DNA 1 in coat 2 and DNA 2 in coat 1.

15-17. The cell culture lyses visibly at 60 min. This time is 20 min (1 latent period) after the moi becomes greater than 1 phage per cell. An approximate outline of the situation can be obtained by assuming that the bacteria divide in steps (see Table A-4). The 100 colony-forming mutants are T7-phage resistant.

15-18. There are several approaches, each of which involves isolating phage mutants that eliminate the activity of X. It is also necessary to have an *in vitro* assay for X; this might be enzymatic or immunologic, or might require isolation of the protein. First, consider a heat-sensitive mutant. It can be shown that at high temperature, enzymatic activity or perhaps immunologic reactivity (not as good a test) is substantially reduced. Second, consider a chain-termination mutation. Here it can be shown by sedimentation or by gel electrophoresis that the protein has a lower molecular weight than the wild-type protein. Note the greater power of the second approach. With the heat-sensitive mutant, one merely correlate a property *in vitro* with that *in vivo*, but with a chain-termination mutant one observes not only the absence of the wild-type protein but the appearance of a new and smaller molecule.

TABLE A-4

Time (min)	Bacteria (per ml)	Phage (per ml)
0	2×10^7	5×10^3
10	—	—
20	—	10^6
30	4×10^7	—
40*	—	2×10^8
50	—	—
60	lysis	—

* All bacteria are infected when concentration of phages is 2×10^8, since moi = 5.

Great elegance can be obtained in the second approach, through which it can be shown that a genetically ordered set of chain-termination mutants makes a set of proteins of graded size. The direction of translation, and therefore of transcription, of the gene is also obtained.

15-20. The sup^+ gene host supports the growth of a phage having a conditional mutation.

15-21. (a) Since the total moi = 5, the fraction of cells which is uninfected is $e^{-5} = 0.007$.
(b) The moi of P2 is 3; the fraction of cells getting no P2 is $e^{-3} = 0.05$. Since 0.007 get neither P2 nor P4, then $0.05 - 0.007 = 0.043$ get one or more P4 but no P2. Hence $0.043 \times 10^8 = 4.3 \times 10^6$ bacteria get P4 but no P2.
(c) Using the above reasoning, $e^{-2} = 0.135$ get no P4, and 0.007 get neither; hence $0.135 - 0.007 = 0.128$, and 1.28×10^7 cells get P2 but no P4.
(d) The number of cells getting at least one P4 and one P2 is $10^8 - (1.28 \times 10^7) - (4.3 \times 10^6) - (7 \times 10^5) = 8.22 \times 10^7$.

15-23. The main point is that the weakening of the cell wall resulting from solubilization of cell-wall components at the site of adsorption is insufficient to cause lysis of the cell. However, unpurified phage stocks contain a substantial amount of phage-induced lysozyme and this can complete the lysis process.

15-25. To prevent incorporation of C into T4 DNA.

15-26. The T4-induced nucleases will destroy all newly synthesized phage DNA molecules since, in cases (a) and (b), glucosylation cannot occur.

15-27. (a), (d), and (e).

15-28. (a), (c), and (d).

15-29. Yes. If the headful rule is followed, the deletion will create terminal redundancy.

15-32. (a) Yes, if different regions of the denaturation map are unambiguously distinguishable.

15-33. The enzymes work catalytically; the structural proteins (of head, tail, tail fibers, and so on) are stoichiometric.

15-37. Some proteins are processed, that is, cut down from larger units.

15-38. From the molecular weight of each known protein, calculate the fraction of the total DNA needed to code for the total molecular weight of the proteins.

15-45. When one λ phage adsorbs to one bacterium, the lytic response ensues and many phages are produced which infect other bacteria. In time the number of phages increases to the point that many bacteria are multiply infected. This favors the lysogenic response. Since the lysogens are immune to subsequent infection by λ, they grow in the plaque and produce a turbid growth of bacteria. Phage T2 lacks a repressor and would not be expected to produce a turbid plaque.

15-46. (a) Repressor in the lysogen prevents transcription of the infecting phage.
(b) Yes, because they would not be recognized by the repressor.

15-47. (a), and (b).

15-49. (b).

15-50. (a) The frequency with which wild-type phages are induced is relatively low, and these form turbid plaques. However, if a mutation occurs in the gene making repressor, the lysogen containing this mutant (which will form a clear plaque) will always produce phages.
(b) The gene *cI* makes repressor. Genes *cII* and *cIII* are needed in the wild-type state for the *establishment* of lysogeny, but not for its maintenance. Thus, a *cII⁻* or *cIII⁻* mutation in a prophage does not result in escape from repression.

15-51. When integrated into *E. coli*, the gene order is

$$gal \quad cI \quad P \quad Q \quad RA \quad J \quad bio,$$

so that *cI* will show the highest cotransduction frequency with *gal*.

15-54. (a) Messenger RNA terminates at the first leftward terminator (t_{L1}) of λ and *N* protein is needed to prevent termination. This conclusion is correct as long as the *trp* operon lacks a promoter.
(b) Yes. If part of gene *N* is deleted, t_{L1} must also be deleted and the mRNA from the main leftward promoter will extend through the *trp* operon.

15-55. Decrease, because *gal* and *bio* are moved apart by the prophage.

15-57. Insertion has occurred at the right prophage *att* site (that is, at PB') to generate a dilysogen. The *ter* system is active and cleaves at the two *cos* sites to produce a J^+P^+ recombinant.

To prove this point, the prophage should be made A^+R^- and the infecting phage A^-R^+. All J^+P^+ phages will be A^+R^+.

15-59. (a) The prophage is transferred to a female lacking repressor; the operators are free of repressor and transcription begins.

(b) Once this locus is transferred into the female, the female cells will die. Hence, by using the interrupted mating technique, one can determine at what time the numbers of *any* recombinant decrease.

15-60. *red⁻*.

15-62. The mutant *ti*12 apparently fails to replicate only when other phages are present; presumably the small burst in the absence of any other phage is also due to a replication defect. It has functioning O and P genes. It is likely that it has a defective replication origin, or, *ori*. This functions weakly when no other phages are present but in a coinfection with a phage having a wild-type *ori*, apparently it cannot compete for the initiation system and replication fails to occur.

15-64. (a) Phage $\phi42$ might be the inverse of λ: it replicates and recombines as a circle but only monomeric DNA can be matured. Thus recombination into a double circle must be followed by a second recombination event to generate the single circle which is the precursor for maturation. This results in a circular map.

(b) For phage λ, the *double circle* is cut at the cohesive sites between A and R to generate the linear monomeric structure found in phage particles. This results in a linear map.

15-65. (a) Repression or establishment of lysogeny.

(b) Complementation between two proteins was needed to establish repression.

15-66. (a). (1) Complementation. (2) In the cI^- infection, only mutant repressor is made, and this mutant does not bind to DNA. In the cI^- and y^- infections, there is no complementation—y^- is a site mutation preventing repressor synthesis by the DNA containing the mutation. (3) $\lambda imm434$ DNA can be used as a control since the cI repressor should cause no binding of this DNA. If one subtracts the binding activity of this phage from the binding activity of the $imm\lambda$ phage, then one should know the $imm\lambda$-repressor binding activity.

(b) The *cro*-constitutive lysogen makes copious amounts of the Cro protein. This protein is seen to repress the cI function of the infecting $imm\lambda$ phage. If a mixed infection is done, then one can see (last row of table) that the $imm434$ phage supplies something to the $imm\lambda$ phage that allows for the expression of the cI-specified proteins in amounts to shut off the Cro protein. It is seen from the data that in this specific case, the Cro protein acts by suppressing the synthesis of proteins by cII and $cIII$ (rows two, three, and four).

15-67. (a) For lytic growth, omit fragments B, C, and E. For growth as a lysogen, omit fragments B and C.

(b) Since the target site of the HindIII endonuclease exhibits two-fold rotational symmetry, the "sticky ends" on each fragment are identical, and therefore some of the fragments will fit in either of two orientations. If fragment E does this, the cI gene is disrupted, so no repressor is made. However, since the fragment is still there, it is possible to reorient it to regain a functional repressor.

(c) (1) A fragment made by using EcoRI could not be substituted, because the "sticky ends" created by EcoRI and HindIII are different and would not anneal. (2) This fragment also would not anneal to the HindIII-generated ends. (3) Since both the λ fragments and the his operon fragment have the same "sticky ends", they could anneal and the hybrid phage could be constructed.

(d) Since the function of the product of gene N is antitermination, then elimination of termination sites t_{R1} and t_{R2} would allow reading of the essential genes in fragment F. None of the other functions in fragment D are needed for lytic growth.

15-68. (a) The E^- lysate contains active tails with tail fiber attached (λ phage has only one tail fiber). These fibered tails can attach to live $E.\ coli$, and the heads from the J^- lysate can attach to preadsorbed tails. Then the λ J^- DNA can inject and infection can proceed, producing J^- progeny.

(b) The genotype is cI^-J^- because this is the DNA contained in the tailless heads.

15-69. The genotype will be $cI857\ P^+A^+$ because the only mode of phage production is Ter-mediated excision at the two cos sites. More phage particles are produced than in problem 15-58 because in problem 15-69 one prophage is P^+ and DNA replication can occur. A typical number of phages is 12 per cell.

15-70. (a) The $groE$ mutation is recessive, since a λ phage carrying the gro^+ allele can multiply in $E.\ coli\ groE^-$.

(b) Phage B is missing the $int\text{-}red$ region of λ, but carries 4,000 bases that include the wild-type allele of the $groE$ gene.

(c) Phage C is phage B with an amber mutation in the $groE$ gene.

(d) Protein X is the product of the $groE$ gene.

(e) Protein Y is an $E.\ coli$ protein whose gene maps near $groE$.

(f) Protein Z is a λ protein coded by a gene in the $int\text{-}red$ region of the λ genome.

15-71. A figure-eight-like structure results which has a single-stranded large loop and a double-stranded small loop. If the λ anneals to an F' strand which is linear owing to a nick in a region other than the prophage DNA, the structure seen will be a small double-stranded loop to which are attached two single strands. The sum of the lengths of the single-strands will equal the length of the single-stranded large circle.

15-73. P22 DNA needs to circularize before it can either lysogenize or undergo more than one round of DNA replication. After infection, generalized recombination is needed for circularization of P22 DNA, by pairing of the homologous, redundant ends. This can be done by the host *rec* system or by the P22 *essential recombination function (erf)*. Once a prophage has been formed by complementation, circularization can be accomplished by the P22 *int* system, and general recombination is no longer necessary. The procedure is explained in Botstein and Matz, *J. Mol. Biol.* (1970, **54:** 417–440).

15-74. The *tet-r* gene, and perhaps other RTF genes, are inserted within the prophage. The prophage excises normally, yielding a phage genome carrying *tet-r;* the genome replicates, but the space inside the phage head is not sufficient to hold all P22 essential genes plus the inserted episomal genes. Thus the phage-like particles produced carry most, but not all, essential genes, and different phage particles lack different essential phage genes. This accounts for their cooperative growth behavior. The high frequency of *tet-r* transduction is due to the fact that most particles will carry the *tet-r* gene. Frequency of *tet-r* transduction cannot be 100 percent, as 90 percent of the infected cells are killed.

15-76. (a) To make the hybrid phage, grow P2 and 186 together in *E. coli* K, and plate the progeny on *E. coli* C(P2). This indicator selects against both parental types and allows the hybrid to plate.
   (b) A host cell of *E. coli* F⁻ with attachment sites for *phe⁻ile⁻* (*h*P2 *imm*186) is constructed and mated with a nonlysogenic *E. coli* Hfr cell with *phe⁺ile⁺* sensitivity. *Phe⁺* and *ile⁺* recombinants are selected. If the *phe⁺* recombinants are mostly cured of the prophage, then the hybrid attaches near *phe*. If the *ile⁺* recombinants are mostly cured of the prophage, then the hybrid attaches near *ile*. One should not use Hfr(*h*P2 *imm*186) and a nonlysogenic F⁻, because the phage might exhibit zygotic induction.
   (c) *E. coli phe⁻ ile⁻ h* P2 *imm* 186 is transduced to *phe⁺* or *ile⁺* with P1 phage grown on prototrophic, nonlysogenic *E. coli*. If the *phe⁺* transductants are 10 percent cured of the prophage, then the hybrid phage attaches near *phe*. If the *ile⁺* transductants are 12 percent cured of the prophage, then attachment is near *ile*.
   (d) The genes involved in prophage attachment are probably closely linked to the immunity gene, not the host range gene, since attachment specificity usually segregates with the immunity gene.

15-78. (a) The injection proceeds in two steps. The first 10 percent of DNA is infected using energy from the phage particle, while the injection of the last 90 percent of the DNA requires cellular energy.
   (b) T5 DNA is unique in sequence (nonpermuted) and one end always injects first.

15-81. The DNA in transducing phages was replicated before infection. Also, either there is no replication of bacterial DNA after infection, or, replicated DNA never gets into transducing particles.

15-83. (a) Complementation.

(b) The *dil* genes are required to set up lysogeny, that is, to establish repression, but are not needed to maintain the lysogenic state. Thus, the *dil* genes code for a positive control element in establishment of lysogeny.

(c) The *tul⁻* lysogens are not stable, since there is no repressor. The *dil⁻* single lysogens can be isolated, since once repression is established (with the help of the *dil* genes of the coinfecting *tul⁻* phage), the *dil* genes are dispensable. The situation is identical to that of λ, if *tul* is *cI* and *dilA* and *dilB* correspond to the *cII* and *cIII* genes.

15-85. This would be possible if they had a double-stranded DNA intermediate which inserted in the chromosome at a defined point.

15-88. The point of this question is that if the phage codes either for its own replicase or initiator for replication, then a successful infection could only occur if the nucleic acid strand was the coding strand. If the origin for replication were on a particular strand, then only that strand would be infectious. Knowing this, you should be able to answer the question.

15-91. Infect spheroplasts with a mixture of φX174 DNA mutant in gene A and one of the fragments. Determine which fragment complements A⁻ mutants. Repeat for each φX174 gene. This method fails for a gene which itself is cleaved by the restriction enzyme.

15-93. Infect a *dnaQ* mutant at the elevated temperature and determine by sucrose gradient sedimentation analysis at what stage DNA replication is blocked.

Chapter 16

16-1. (a) *leu⁻pro⁻*.   (b) Colonies on Leu and Pro plates are *leu⁺* and *pro⁺* revertants, respectively. Thymine is not required since reversion for two markers would be unlikely. No colonies on Thy plate for these would be double revertants, that is, *leu⁺pro⁺*.

16-2. (1) For all markers, +. (2) *thr⁻*, others +; except for *pro*, which is unknown. (3) *arg⁻, pro⁻*, others +.

16-3. No. There is nothing with which to complement.

16-4. *bac.*

16-5. Cotransduction frequencies are: *pur-pro*, 25/286 = 9 percent; *pur-his*, 159/286 = 56 percent; *pro-his*, 1/286, < 1 percent. The gene order must be *pro-pur-his*, or *his-pur-pro*, depending on how you look at it. The *pur-his* distance must be much shorter than the *pur-pro* distance.

16-6. The original strain S had a *lac⁻* point mutation in the chromosome and contained a sex factor F' Lac, which is temperature-sensitive, in that it fails to replicate at 42° C. Variant 1 contains a *ts⁺* revertant of F' Lac; 2

is a lac⁺ recombinant; 3 has F′ Lac inserted into the gene for T1 sensitivity.

16-8. (a) Statement 2.

(b) Yes, but not until the time when prophage λ enters, at which point there will begin a steady decrease with time of the number of recombinants.

(c) Yes, since there will be no prophage induction.

16-11. *Pro* is a terminal marker. F is closely linked to Pro (that is, all *pro⁺* recombinants are donors and hence *F⁺*) and is therefore also at the terminus.

16-12. The direction depends upon the orientation of the inserted F.

16-14. Transfer is by way of a rolling circle mechanism. After *z* is transferred, *a* is retransfered; hence *z* and *a* are closely linked.

16-15. The *arg⁻* mutants have different map locations and must be in different genes; one is transferred before *met*, the other is transferred after *met*.

16-16. (a) Six genes.

(b) Some of the closely linked mutations might lie in closely linked genes, that is, they might complement one another. Moreover, there might be other genes in the pathway for which no mutants have been isolated. Ten mutations distributed according to Poisson's law over seven or eight genes would stand a good chance of leaving one gene untouched.

16-17. Between *c* and *d*.

16-18. (a) True, because *a* enters before *c*.

(b) False, because *b* enters before *c*.

(c) True, because some crossovers will be between *a* and *b*.

(d) False. See (c).

(e) True, because *b* is between *a* and *c*.

(f) True, because the *a⁻* and *b⁻* alleles are near, and on, the same DNA.

(g) True, because *b* enters after *a*.

16-19. Both (a) and (d) are possibilities. In (a) one would measure the survival of female recombinants as a consequence of zygotic induction. In (d) one could study loss of λ immunity by recombination, but this would be tedious.

16-21. (a) The genotype is *his-ts trp⁻ leu⁻*.

(b) The strain is *his-ts* and does not require histidine at 25° C.

(c) The leucine requirement results from a deletion or double mutation.

(d) The genotype is *his-ts leu⁻ trp⁺*.

(e) The genotype is *leu⁻ trp⁻ his⁺*.

(f) (1) *his, leu, trp*, in that order. (2) Perform an interrupted mating. (3) *Met⁻* is the counterselective marker; it allows the male bacteria to be selected against.

16-22. (a) The gene order is *galA galB galC bio*.

(b) $C > B > A$, because nonsense mutations are the most likely to give a Gal$^-$ phenotype for each mutational event (that is, each base-change).

16-24. (a) Reciprocal recombinant classes are produced in equal amounts in two factor crosses independent of the degree of equality of parental input; therefore, 7 percent $h^- m^-$.

(b) More of the interactions will be between phage of opposite parental genotype when the inputs are equal, so *more* wild-type recombinants will arise.

(c) The frequency of recombinants depends on interactions between phage of opposite parental type, so it depends on the product $w(1 - w)$ where w is the fraction of parental phages having the minority genotype. Thus

$$\frac{\text{frequency of wild-type in cross (b)}}{\text{frequency of wild-type in cross (a)}} = \frac{0.5 \times 0.5}{0.15 \times 0.85},$$

which yields 14 percent wild-type recombinants in (b).

16-25. In the Visconti-Delbrück theory,

$$R = 2w \ (1 - w) \ (1 - e^{-mp}),$$

in which R is the recombination frequency, w is the fraction of parental phages having the minority genotype, m is the number of rounds of mating, and p is the probability of recombination per mating. For genes unlinked in the Mendelian sense, $p = 1/2$. In standard crosses, $w = 1/2$. Hence

$$R = 1/2 \ (1 - e^{-m/2}) = 0.12 \text{ and } m = 0.54.$$

16-26. The vegetative map is

$$\underline{a \quad d \quad c \quad b.}$$

The prophage map is

$$\underline{lac \quad c \quad b \quad a \quad d.}$$

The two sets of data suggest that a linear, unique DNA molecule forms a circle which inserts by recombination between c and d. From known phages, the simplest guess for the DNA structure is one with complementary single-stranded ends. This might be tested by circle formation or end-to-end aggregation *in vitro*, as judged by sedimentation or electron microscopy.

16-29. The new *rII* mutation must be a deletion mutation which spans the gap between the A and B cistrons. It will eliminate parts of both genes but not the parts which contain the indicated point mutations.

16-30. (a) Complementation.

(b) Three cistrons $(c1, c4)$, $(c2, c3)$, and $(c5)$.

(c) No. Complementation does not make a map—at least not an arrangement of the genes relative to one another.

(d) Phage mutant $c5$ must define a gene required to maintain lysogeny, since no lysogenic strains carrying $c5$ are ever observed. The other mutants yield lysogenic strains at low frequency, so they must be able to maintain lysogeny once it has been established.

(e) Clear mutants do not allow the temperate phage a choice between the lytic and lysogenic pathways of development, so no lysogenic bacteria ever grow up in the plaque centers.

(f) A nonhomologous temperate phage may either lyse the cell, since it is not repressed for the unrelated phage, or it may undergo normal lysogenization, in which case the bacterial strain will become doubly lysogenic.

16-31. (a) Mutant $ts1$ is most leaky.

(b) Five cistrons are defined, since $am1$ and $ts4$ are in the same complementation group, as are $ts5$ and $ts6$.

(c) $am1 \xleftrightarrow{\hspace{1cm} 5\% \hspace{1cm}} \begin{array}{c} ts2 \\ ts1 \end{array} \xleftrightarrow{\hspace{1cm} 5\% \hspace{1cm}} ts3 \xleftrightarrow{\hspace{1cm} 5\% \hspace{1cm}} ts5$

(d) To get the relative order of $ts1$ and $ts2$, construct $am1\ ts1$ and $am1\ ts2$. Perform the two crosses:

$$am1\ ts1 \times ts2$$

and

$$am1\ ts2 \times ts1$$

in a sup^+ host at 30°. Plate the progeny on the same host at 42°, to select for ts^+ recombinants, while keeping $am1$ as a nonselective marker. The percent of am phage among the ts^+ recombinants ought to show the relative order of $ts1$ and $ts2$.

16-32. (a)

$$\overset{\text{2.5}\hspace{1cm}\text{2.1}\hspace{1.5cm}\text{5.0}}{\underset{h\hspace{1.2cm}(am1,\,ts2)\hspace{0.5cm}ts3\hspace{2cm}ts1}{\rule{7cm}{0.4pt}}}$$

Order and spacing
is uncertain

Distances can be uncertain to ±0.1 map units.

(b) h $am1$

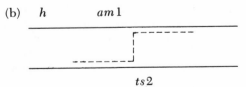

$ts2$

Not as follows:

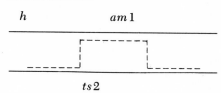

The second cross cannot resolve the ambiguity, since a majority of recombinants would be expected to carry the *am*1 mutation, regardless of the order of *am*1 and *ts*2. To wit:

or

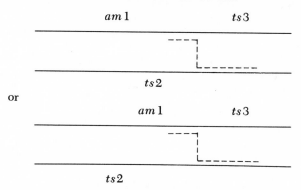

16-33. (a) Deletions *r*A105, *r*638, *r*PB230, and *r*1993.

16-34. Addition of two bases or deletion of one base.

16-35. (a) Mate 1 and 2 for a short period of time and select a *gal*⁺ *str-r* recombinant (that is, a fourth strain). Mate 3 with 4 for a short time and select for *gal*⁻ *str-r* recombinants. Streak these on a plate lacking proline and pick a *pro*⁺ clone.
(b) To make a *str-s* strain is difficult with these strains since if the *str-s* allele is transferred to the *F*⁻ strain, then there is no way to select against the Hfr strain. This can be done if an additional marker is added. For instance, the Hfr could be made auxotrophic for a gene transferred *later* than the *str* locus and the mating mixture could be plated on agar lacking this nutrient. Alternatively the *F*⁻ could be made resistant to a phage or another antibiotic. These markers should be far from the *str* and *gal* loci though.

16-36. (a) Cross *bio*11 *c*⁺ with *cI*857. Plate on Q5175 at 42° C and pick a clear plaque.
(e) Cross *c*⁺ with *cI*857 *P*⁻. Plate on Q5151 at 42° C and pick a turbid plaque.

16-38. (a) Grow the *F*⁺ culture on agar containing acridine orange. Test colonies either for inability to transfer markers, ability to be a recipient, or resistance to male-specific phages.

16-41. An outline of the procedure is the following: Make *bio* 11 *H⁻* and cross with *xis* 1 selecting against the *bio* locus. Put the *cI* markers in the appropriate places, using them to distinguish parents from recombinants. Test for the *H* character by ability to grow on a *sup⁻* host.

16-43. (a) The cotransduction frequencies order the mutants with respect to the *uvrC* gene, higher frequencies being nearer. See (b) for position of 4.

(b) In the three-factor cross 4 cannot recombine with 2 or 3. The simplest explanation is that 4 is a deletion and the region deleted contains the sites of the mutations 2 and 3. The complementation results define the genes. That is, 1, 3, and 5 mutually complement and therefore each must be in a separate gene.

(c) No, mutants which recombine do not necessarily complement each other. Recombination can occur between single nucleotides, and genes tend to be several hundred nucleotides long. Mutant 1 fails to complement 2, even though they are mutant in separate genes. Hence, 2 is a polar mutation, consistent with its characterization as a nonsense mutant. Thus, the lack of complementation between 1 and 2 need not mean that they are mutant in the same gene. The cotransduction frequencies give an idea of gene size; thus either mutant 2 does occur in a gene separate from 3 or the gene containing 2 and 3 is very large. Mutant 4 is a polar mutation affecting 1, and might be so because a shift in the reading frame resulted in a nonsense signal which was not read in the wild-type but was read in the deletion mutant 4.

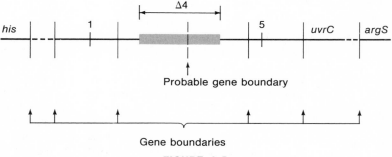

FIGURE A-5

16-45. Prophage integration and excision; formation of a heterogenote; formation of an F′ factor from an Hfr; and formation of an Hfr from an *F⁺* cell.

16-47. The frequency would be higher when an inhibitor of DNA synthesis is added, because break–rejoin heterozygotes are all internal heterozygotes and are therefore destroyed by replication.

16-50. Configurations 3 and 5.

Chapter 17

17-1. The periods of the cell cycle are G1, S (DNA synthesis), G2, and M (mitosis), in which G1 and G2 refer only to the two intervals between S and M.

17-3. C_0t analysis shows that there are fractions of the DNA which renature at different rates. The theory of C_0t analysis enables one to calculate the relative number of copies contained in each fraction from the relative rates of renaturation.

17-6. Cells which in nature grow in the form of tissues grow in cell culture until the cells are in contact with one another. At this point they have formed a monolayer and stop growing. Tumor cells continue to grow and form disorganized, piled-up masses of cells.

17-8. Transformed cells do not show density-dependent growth or anchorage dependence whereas normal cells do. They require less serum for growth and are more easily agglutinated by lectins than normal cells. Furthermore, they frequently excrete plasminogen activator. A transformed cell is presumably a prototype for a tumor cell.

17-10. Quinacrine is a fluorescent dye which binds to chromosomes, producing a pattern of fluorescent bands characteristic of a particular chromosome. It is used to identify chromosomes which are otherwise morphologically indistinguishable.

17-12. These cause destruction of the mitotic spindle; thus they produce a block in metaphase.

17-15. $28s$ and $18s$ RNA molecules are found in $60s$ and $40s$ ribosomes, respectively.

17-16. Penicillin interferes with bacterial cell wall synthesis and therefore has no effect on eukaryotes. Actinomycin affects transcription in *both* bacteria and mammals and therefore is not of clinical usefulness. The remainder affect translation, but tetracycline and erythromycin cannot inhibit translation in mammalian cells. These two are useful against infection. Puromycin and chloromycetin are active against both bacterial and mammalian ribosomes. However, in cases in which an infection is becoming dangerous to a patient, chloromycetin can be used but the toxic effects have to be dealt with in other ways.

17-19. Usually from membranes in the cell.

17-23. Supercoiled DNA. Occasionally if the DNA is not isolated carefully, open circles and even linear molecules can be found.

17-25. The enzyme reverse transcriptase results in the synthesis of double-stranded DNA from an RNA template and this DNA can be inserted into the chromosome of the host.

Index of Problems

Index of Subjects